The City & Guilds textbook

Plastering

LEVEL 1 DIPLOMA (6708)
LEVEL 2 DIPLOMA (6708)
LEVEL 2 TECHNICAL CERTIFICATE (7908)

Mike Gashe
Kevin Byrne

Orders: please contact Hachette UK Distribution, Hely Hutchinson Centre, Milton Road, Didcot, Oxfordshire, OX11 7HH. Telephone: +44 (0)1235 827827. Email education@hachette.co.uk Lines are open from 9 a.m. to 5 p.m., Monday to Friday. You can also order through our website: www.hoddereducation.co.uk

ISBN: 978 1 3983 0647 9

© The City & Guilds of London Institute and Hodder & Stoughton Limited 2020

First published in 2020 by
Hodder Education,
An Hachette UK Company
Carmelite House
50 Victoria Embankment
London EC4Y 0DZ

www.hoddereducation.co.uk

Impression number   10 9 8 7 6 5 4 3 2

Year   2024 2023

Cover photo © ahavelaar - stock.adobe.com

City & Guilds and the City & Guilds logo are trademarks of The City and Guilds of London Institute. City & Guilds Logo © City & Guilds 2020

Illustrations by Integra Software Services Pvt Ltd

Typeset by Integra Software Services Pvt. Ltd., Pondicherry, India

Printed and bound by CPI Group (UK) Ltd, Croydon CR0 4YY

A catalogue record for this title is available from the British Library.

Contents

Acknowledgements

This book draws on several earlier books that were published by City & Guilds, and we acknowledge and thank the writers of those books:

- Mike Gashe
- Michael Mann
- Colin Fearn
- Martin Burdfield

We would also like to thank everyone who has contributed to City & Guilds photoshoots. In particular, thanks to: Andrew Buckle (photographer), Mike Gashe and the staff and students at Grŵp Llandrillo Menai, Michael Mann and the staff and students at South and City College Birmingham.

Contains public sector information licensed under the Open Government Licence v3.0.

From the authors

My contribution to this book is dedicated to my wife Gail Byrne, who has supported me throughout this process, and to my recently late daughter, Nicola Ann Byrne, who would have been immensely proud of my efforts.

Kevin Byrne

Thanks to Emma and Matt and the rest of the team at Hodder Education for allowing me the opportunity to develop the new Plastering textbook. Thanks also to City & Guilds for the material in the previous editions, and to British Gypsum for the use of their literature and images.

Mike Gashe

About the authors

Kevin Byrne

I was born and brought up in Oldham, Lancashire, where I have lived and worked all my life. I come from a family of plasterers: grandfather, great uncle, father, brother, myself and my son, so was essentially born into the trade. I worked my school holidays from thirteen years old helping my father and brother. In 1977, I left school at the age of fifteen and got an apprenticeship in plastering at Oldham Council. I didn`t enjoy this because I was working mainly on small repairs to tenanted houses, however, I soon went to work with my father and brother on housing and commercial work, which was tough, but I enjoyed it immensely. My brother and I worked together for five years from 1982 before I started my own successful business, sometimes employing up to 50 people depending on the size of the contracts. Although for a large amount of that time I had no formal qualifications I became a competent solid plasterer and businessman. In 2007, I was asked to help out at The Oldham College teaching practical plastering to adults. While teaching at The Oldham College I gained my L2 and 3 NVQ in solid plastering and also gained my teaching qualifications: PTLLS 7303, Certificate in Education, Assessor Awards and BA in Education and Professional Development. I stayed at The Oldham College teaching on a part-time basis until 2011 and continued to run a scaled-down version of my plastering business, employing myself and two operatives. In 2013, I was offered an opportunity to teach plastering full time at Kirklees College, which I took and am currently still there. While there I gained my Internal Verification Awards. I am a member of the National Association of Plastering Lecturers and an Associate APL of the Worshipful Company of Plaisterers. I currently do consultancy work for City & Guilds developing plastering qualifications. I find my current roles in teaching and developing immensely satisfying and take great pride in influencing and developing the next generation of plasterers in this brilliant trade, which is a highly skilful and tough environment to work in but has given me the opportunity to earn a good living and enabled me to support my family throughout my working life. In my spare time I have always kept fit and when younger played Rugby League and Squash to a good amateur competitive level. I am also a keen cyclist, both road and mountain biking, when time allows, and avidly follow all these sports live and on TV.

Mike Gashe

I was brought up in a small village in the Snowdonia National Park, North Wales where I continue to live with my family. My plastering career started at the age of 16, attending Bangor Technical College under a three-year apprentice scheme. During my time as a student at the college I won the Blue Circle Apprentice Award as well as the British Gypsum Young Apprentice Plasterer Award.

Following my time at the college I went on to work as a self-employed commercial and private plastering contractor for a number of years, and today I continue to practise those traditional and modern skills and techniques on a smaller scale.

For the past 26 years I have been a lecturer and programme leader in the plastering section at Grwp Llandrillo Menai. I'm proud and honoured that during my time at the college many of our learners, including the plastering department, have been successful in achieving Regional and National awards.

During the last 9 years I've also worked closely with City & Guilds, developing and writing new qualifications and assessment material, including developing smart screen teaching resources and training manuals for plastering and dry lining to support teaching staff. I have recently become a Principal Moderator and chief examiner for City & Guilds technical qualifications.

Publishing this book has given me the opportunity to reflect on my own plastering experiences and be able to pass on this knowledge to other learners to help them achieve their future aims.

Kevin and Mike would like to add that if you talk the talk and walk the walk success and rewards will come your way.

Picture credits

All step-by-step sequences © City & Guilds 2020, unless otherwise credited.

p.6 t © Ewelina Wachala/Shutterstock.com, m © Ewelina Wachala/Shutterstock.com, b © Esdras700/stock.adobe.com; Fig 1.2 © Megis/stock.adobe.com; Fig 1.3 © Andrey Demkin/Shutterstock.com; Fig 1.5 © Cheng Wei/Shutterstock.com; Fig 1.7 © Mr.Arthid Vongsawan/Shutterstock.com; Fig 1.9 © Trongnguyen/stock.adobe.com; Fig 1.25 Image courtesy of Ayrshire Metals Ltd; Fig 1.29 © OneClic/stock.adobe.com; Fig 1.30 © Silvio/stock.adobe.com; Fig 1.31 © Travelwitness/stock.adobe.com; p.21 © Simon/stock.adobe.com; p.22 1st © Yoki5270/Shutterstock.com, 2nd © Peeradontax/stock.adobe.com, 3rd © Oleksii/stock.adobe.com; Fig 1.32 © Ronstik/stock.adobe.com; Fig 1.33 © Soleg/stock.adobe.com; Fig 1.34 © Never Paint Again UK; Fig 1.36 © Chassenet/BSIP SA/Alamy Stock Photo; Fig 1.37 © Hcast/stock.adobe.com; Fig 1.38 © CorkSol UK; Fig 1.39 © British Gypsum; Fig 1.40 © Kris Mercer/Alamy Stock Photo; Fig 1.41 © Namis Development Ltd; Fig 1.42 © Giordano ravazzini/Shutterstock.com; Fig 1.43 © Dusan Kostic/stock.adobe.com; Fig 1.44 © Varvara Gorbash/123 RF.com; Fig 1.46 © Ambiance chaleur/stock.adobe.com; Fig 1.47 © Sinn P. Photography/Shutterstock.com; Fig 1.48 © Wideonet/stock.adobe.com; Fig 1.51 © Ron Hudson 2008/Fotolia; Fig 1.52 © Chad McDermott/Shutterstock.com; Fig 1.53 © Robert Kneschke/stock.adobe.com; Fig 1.54 © Vaceslav Romanov/stock.adobe.com; Fig 1.55 © Ungvar/stock.adobe.com; Fig 1.56 © MPS Electrical Ltd; Fig 1.57 © Imagine/stock.adobe.com; Fig 1.58 © Photographee.eu/stock.adobe.com; Fig 1.59 © AJC Carpentry; Fig 1.60 © Ungvar/stock.adobe.com; Fig 1.61 © Toat/stock.adobe.com; Fig 1.62 © RTimages/stock.adobe.com; Fig 1.63 © Andrey_Popov/Shutterstock.com; Fig 1.68 © Murattellioglu/stock.adobe.com; Fig 1.70 © Torsakarin/stock.adobe.com; Fig 1.71 © Jameschipper/stock.adobe.com; Fig 1.72 © Zlatko Guzmic/stock.adobe.com; Fig 1.74 © ZZTop1958/Shutterstock.com; Fig 1.75 © Tasha and Deki/Shutterstock.com; Fig 1.76 © Morphart/stock.adobe.com; Fig 1.84 © Richard Johnson/stock.adobe.com; Fig 1.86 © Image'in/stock.adobe.com; Fig 1.87 © John Price/Alamy Stock Photo; Fig 1.88 © Portakabin;

p.52 © Chokniti/stock.adobe.com; p.53 t © JeanLuc/stock.adobe.com; m © R_Yosha/stock.adobe.com; b © Christian Schittich/Alamy Stock Photo; p.54 t © David Stone; b © Idmanjoe/stock.adobe.com; p.55 t © Franco lucato/stock.adobe.com; b © FluxFactory/E+/Getty Images; p.56 t © Hansenn/stock.adobe.com; p.57 t © Alan/stock.adobe.com; m © Alexandr Mitiuc/stock.adobe.com; b © Dantok/stock.adobe.com; p.58 t © Majeczka/stock.adobe.com; Fig 2.2 © A/stock.adobe.com; Fig 2.3 © Virynja/stock.adobe.com; Fig 2.6a © Rigsbyphoto/Shutterstock.com; Fig 2.6b © David J. Green - electrical/Alamy Stock Photo; Fig 2.6c © Ornot38/Shutterstock.com; Fig 2.7 © Rob Kints/Shutterstock.com; p.70 t © Rosamar/Shutterstock.com, m © James Hughes/Alamy Stock Photo, b Reprinted courtesy of Engineering News-Record, copyright BNP Media, July 11, 2018, all rights reserved.; p.71 1st © D.Semra/Shutterstock.com, 2nd © photomelon/stock.adobe.com, 3rd © DenisNata/Shutterstock.com, 4th © A/stock.adobe.com, 5th © Petrik/stock.adobe.com; p.72 t © Stockbyte/Getty Images/Entertainment & Leisure CD35, m © modustollens/stock.adobe.com, b © Dorling Kindersley ltd/Alamy Stock Photo; Fig 2.8 © Phovoir/Shutterstock.com; Fig 2.9 © 2dmolier/stock.adobe.com; Fig 2.10 © Goodluz - Fotolia.com; Fig 2.11 © HSE; Fig 2.12 © Dmitry Kalinovsky/Shutterstock.com; Fig 2.13 © Mediscan/Alamy Stock Photo; Fig 2.14 © BuildPix/Avalon/Construction Photography/Alamy Stock Photo; Fig 2.15 © Mark Richardson/Alamy Stock Photo; Fig 2.16a © Mr.Zach/Shutterstock.com; Fig 2.16b © Mark Sykes/Alamy Stock Photo; Fig 2.16c © Science Photo Library/Alamy Stock Photo; Fig 2.16d © Colin Underhill/Alamy Stock Photo; Fig 2.17 © Coolvectormaker/stock.adobe.com; Fig 2.18 © Icedea/stock.adobe.com; Fig 2.19 © ipm/Alamy Stock Photo; Fig 2.20 © 29september/Shutterstock.com; Fig 2.21 © Mint Images Limited/Alamy Stock Photo; Fig 2.22 © John Price/Alamy Stock Photo; Fig 2.23 © Steroplast Healthcare Limited; p.91 1st © Dorling Kindersley ltd/Alamy Stock Photo, 2nd © Rollins and Sons Ltd, 3rd © Bioraven/stock.adobe.com, 4th © Gelpi/stock.adobe.com; p.92 1st © City & Guilds, 2nd © 2020 Screwfix Direct Limited, 3rd © 2020 Screwfix Direct Limited, 4th © Refina/SUPERFLEX, 5th © Stanley Black and Decker, 6th © Stanley Black and Decker, 7th © Alexander/stock.

How to use this book

Throughout this book you will see the following features.

Industry tips are particularly useful pieces of advice that can assist you in your workplace or help you remember something important.

INDUSTRY TIP

In the apprenticeship training period of your career, you will be paid an agreed amount per week. When fully competent, plasterers are usually paid per square metre of materials applied to surfaces.

Key terms in bold purple in the text are explained in the margin to aid your understanding. (They are also explained in the Glossary at the back of the book.)

KEY TERM

Corrosion: any process involving the deterioration or degradation of metal components, where the metal's molecular structure breaks down irreparably.

Health and safety boxes flag important points to keep yourself, colleagues and clients safe in the workplace. They also link to sections in the health and safety chapter for you to recap learning.

HEALTH AND SAFETY

A fire extinguisher should always be available when using any form of soldering equipment.

Activities help to test your understanding and encourage you to learn from your colleagues' experiences.

ACTIVITY

What would motivate you to improve your work? Make a note and discuss with your team to see what motivates them.

Improve your maths features provide opportunities to practise or improve your maths skills.

Improve your English features provide opportunities to practise or improve your English skills.

At the end of each chapter there are some **Test your knowledge** questions. These are designed to identify any areas where you might need further training or revision.

The **apprenticeship only** icon identifies content that is relevant to apprenticeship learners only.

CHAPTER 1

PRINCIPLES OF CONSTRUCTION

INTRODUCTION

There is much more to work in the construction sector than just individual trade areas. There are many generic areas which fit into all parts of the construction process. For the work to be of a good standard and cost-effective, it needs to be well-organised and efficient.

The construction industry covers many different areas such as domestic new build houses and commercial new build such as hospitals, schools, factories, roads, bridges, prisons and train lines. There is also the renovation of dilapidated buildings, including heritage work on listed buildings, and facilities work such as the upkeep and repair of existing buildings.

To work in construction, you will need to be able to read and interpret construction drawings. You need to have a good understanding of maths to work out calculations and quantities of materials and you must be able to communicate positively. By reading this chapter you will gain knowledge of:

1. how to understand the construction industry
2. different types of construction information
3. building substructure: foundation construction
4. building superstructure: floor, wall and roof construction, with internal finishes
5. how to interpret building information
6. good practice for setting up the site
7. how to communicate at all levels of the process
8. how to understand and be aware of current environmental good practice within the construction process.

The table below shows how the main headings in this chapter cover the learning outcomes for each qualification specification.

Chapter section	Level 1 Diploma in Plastering (6708-13) Unit 101/501	Level 2 Diploma in Plastering (6708-23) Unit 202/602	Level 2 Technical Certificate in Plastering (7908-20) Unit 201
Understand the construction industry			Topics 1.1, 1.2
Understand construction information			Topic 2.1
Building substructure	Learning outcome 3	Learning outcome 12	Topics 4.1, 4.2, 4.3
Building superstructure	Learning outcomes 4, 5, 6	Learning outcomes 13, 14, 15	Topics 5.1, 5.2, 5.3, 5.4, 5.5
Interpret building information	Learning outcome 1	Learning outcome 10	Topics 2.2, 2.3
Setting up and securing work areas			Topics 2.4, 3.1, 3.2
Communication	Learning outcome 7	Learning outcome 16	Topic 1.3
Sustainability	Learning outcome 2	Learning outcome 11	

1 UNDERSTAND THE CONSTRUCTION INDUSTRY

Areas of construction work and types of organisation

There are many different areas of construction work, which are explained in this table.

Types of construction	Definition
New build	New construction, rather than renovation or refurbishment of existing structures. New build can include all types of construction such as houses, apartments, office blocks, hospitals and stadiums.
Renovation	Improvements to a broken, damaged or outdated building. Usually this will be commercial or residential property. It could be described as bringing something back to life, perhaps to be used for a different purpose, such as an old cinema changed to a restaurant/leisure facility.
Maintenance	Making sure buildings continue to look good and operate at optimum efficiency. If a building is not properly maintained, it can become inefficient. This will make it more expensive to operate and might affect the health and safety of users. A good example of maintaining a building is changing all light bulbs to energy-saving efficient ones.
Restoration	The process of returning a historic building to its former state. The aim will be to accurately recreate the building's form, features and character as they appeared at a particular time. Traditional materials and techniques might be used to reflect the time period that the restoration is aiming to recreate.
Domestic	Projects involving extensions, repairs and refurbishment work on the homes of domestic clients. These are usually undertaken by small building contractors.
Commercial	Building projects such as offices, industrial factories and business establishments.
Industrial	Buildings for larger businesses: steel-erected industrial buildings, such as warehouses and manufacturing plants.

Many different types of organisation are involved in the construction process.

Organisation	Definition
Building contractors	Contractors co-ordinate the resources needed to carry out different types of construction work. This might include contracting a number of trades needed to carry out the works, such as plasterers, bricklayers, joiners, plumbers and electricians.
Manufacturers	An organisation or business that makes goods to sell. Manufacturers take raw materials or components from different sources, assemblies and other materials and turn them into finished products which can then be sold to customers. British Gypsum is an example of a business that manufactures and supplies gypsum-based materials for plasterers to use in construction.
Suppliers	Suppliers provide products or services to others. They often receive products directly from the manufacturers and then distribute them to other entities at a profit. For example, builders' merchants obtain materials from manufacturers and then sell and supply to their own customers for a percentage profit.
Local authorities	Local authorities provide housing in a particular area and are also responsible for property maintenance. They need to assess the housing needs for the area and build, buy and lease dwellings to match those needs. They might also provide loans for repairing and improving dwellings.
Legislative bodies	These bodies prepare and enact laws that have been passed by Parliament. Bodies such as the Health and Safety Executive (HSE) and Building Control ensure the construction industry is upholding the law.

HEALTH AND SAFETY

The HSE governs health and safety law, quality and standards. For more on the HSE, see Chapter 2.

IMPROVE YOUR ENGLISH

'Optimum' means 'best'. Can you construct a plastering-related sentence using the word 'optimum'?

The construction industry is made up of many companies and types of business offering different services, such as plumbing and electrical. Companies can be small (1–49 employees), medium (50–249 employees) or large (250 employees and above). Depending on the client, these companies work in two different sectors.

1 Public: schools, hospitals, libraries and public buildings are generally financed by government/ public funding.
2 Private: individuals or groups of people/consortiums fund work, from a small extension to a huge private housing development.

ACTIVITY

1 Work with a partner to find examples of all three categories of different-sized company.
2 Find examples of a construction company with only one employee (**sole trader**).

Roles of team members and career progression

Jobs and roles in the construction industry are defined in four different sectors.

1 Building: physical construction/making of a building, including maintenance, restoration and refurbishment.
2 Civil engineering: the construction and maintenance of roads, railways, bridges and sewers.
3 Electrical engineering: installation and maintenance of electrical systems including lights, power sockets and appliances.
4 Mechanical engineering: installation and maintenance of heating, ventilation and lift systems.

Employment within construction is in four areas:
- professionals
- technicians
- building craft workers
- building operatives.

Professionals

Professionals are usually educated to degree level.

Architect

Architects design new buildings and the redevelopment of existing buildings. They are also involved in finding ways to restore and maintain old buildings. Architects will be involved throughout a building project from the initial planning stage until the work is completed.

Structural engineer

Structural engineers ensure buildings and structures are designed safely and existing buildings are structurally secure so they can withstand the different elements to which they will be exposed. Their work can include designing and working out calculations for steel structures. Their work is essential for obtaining planning permission and building regulations approval.

Surveyor

Surveyors offer advice on many aspects of design and construction, including maintenance, repair, refurbishment and restoration of proposed and existing buildings. They offer quality assessments and report on defects in or ways of improving all kinds of building. They perform searches for types of land and its suitability for building upon.

Service engineer

Service engineers give advice on the design, installation and maintenance of buildings' services such as water, lighting, heating, air conditioning, lifts and telecoms/ Wi-Fi. They will look for the most cost-effective and energy-efficient systems.

Technicians

Technicians link professional workers with craft workers. There are many different types of technician.

Architectural technician

The role of an architectural technician is to provide architectural design services and to work closely with building professionals on construction projects. They assess the needs of clients and users and agree the **project brief**.

KEY TERM

Sole trader: a self-employed person who owns and runs their own business. The business does not have a legal identity separate to its owner, so that person is the business.

Project brief: a summary of a project's ideas; it shows what work needs to be done.

Building technician

A building technician supports construction managers, surveyors, architects and other workers on construction sites. Their job involves estimating material costs, negotiating the price of and buying materials and checking the quality of suppliers.

Quantity surveyor

A quantity surveyor manages the costs of a project, making sure it is completed within the planned budget. They are responsible for the contractual and financial side of the work.

Building craft workers

Craft workers are skilled people who work with materials to physically construct buildings. There are many different types of craft worker.

Plasterer

A plasterer applies layers of plaster and plasterboard to internal and external walls and ceilings, to provide a protective surface ready for decoration.

Carpenter/Joiner

A carpenter or joiner works with wood to create different fixtures, fittings and constructions. Joiners usually work in a workshop setting (such as making windows and staircases), while carpenters usually install materials on a construction site (such as fixing architraves and skirting boards).

Bricklayer

A bricklayer constructs the first shell of a building using bricks, blocks and mortar. They set out the buildings in line with approved construction plans.

Painter and decorator

These craft workers prepare and decorate internal and external surfaces to match the client's wishes. They need to be able to work on a variety of surfaces, including metal, wood, plaster and render, and use materials such as paint, varnish and wallpaper.

Electrician

Electricians install, maintain and repair electrical control, wiring and lighting systems and read technical diagrams. They perform general electrical maintenance and inspect transformers, circuit breakers and other electrical components.

Plumber

Plumbers work with gas and water systems, mapping the layout for pipes, drainage systems and other plumbing materials, based on the building specifications. They install pipes, sinks, toilets, baths, showers and any other fixtures related to the plumbing of the structure.

Slater and tiler

Slaters and tilers ensure the roof system is watertight. This might include work on flat roofs and pitched roofs (see page 27). They use different methods for slating and tiling, sometimes using traditional methods for restoration work, or synthetic slates/tiles, felt sheets, waterproof bitumen or liquid fibreglass systems for modern buildings.

Woodwork machinist

A woodwork machinist cuts and prepares timber for builder's merchants, DIY stores and for use in furniture-making and construction industries. The timber that they prepare might be used for floorboards, staircases, kitchen units, bars and cabinets. To cut and shape the timber, they use lathes, bench saws, planers and sanders.

Building operatives

There are two types of operative in construction.

Specialist building operative

This type of operative carries out specialist activities such as dry walling, asphalting, scaffolding, floor and wall tiling and window fitting.

General building operative

This type of operative is commonly known as a ground worker and carries out non-specialist operations such as kerb laying, concreting, path/ flag laying and drainage work, supported by general labourers. They might be involved in digging and laying foundations, using plant and machinery such as digger's drills and pumps.

Figure 1.1 shows the different people who make up a building team.

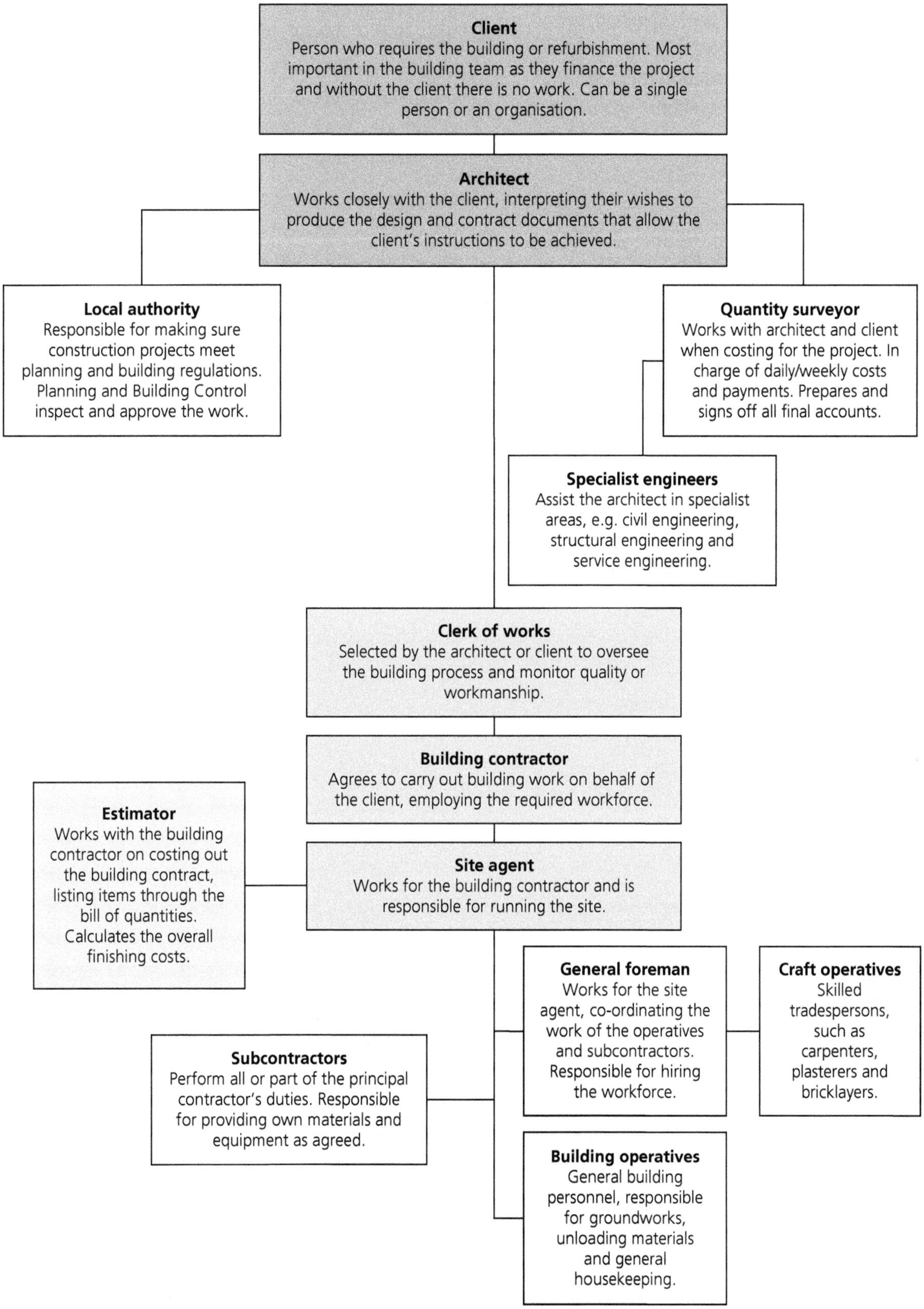

Client
Person who requires the building or refurbishment. Most important in the building team as they finance the project and without the client there is no work. Can be a single person or an organisation.

Architect
Works closely with the client, interpreting their wishes to produce the design and contract documents that allow the client's instructions to be achieved.

Local authority
Responsible for making sure construction projects meet planning and building regulations. Planning and Building Control inspect and approve the work.

Quantity surveyor
Works with architect and client when costing for the project. In charge of daily/weekly costs and payments. Prepares and signs off all final accounts.

Specialist engineers
Assist the architect in specialist areas, e.g. civil engineering, structural engineering and service engineering.

Clerk of works
Selected by the architect or client to oversee the building process and monitor quality or workmanship.

Building contractor
Agrees to carry out building work on behalf of the client, employing the required workforce.

Estimator
Works with the building contractor on costing out the building contract, listing items through the bill of quantities. Calculates the overall finishing costs.

Site agent
Works for the building contractor and is responsible for running the site.

General foreman
Works for the site agent, co-ordinating the work of the operatives and subcontractors. Responsible for hiring the workforce.

Craft operatives
Skilled tradespersons, such as carpenters, plasterers and bricklayers.

Subcontractors
Perform all or part of the principal contractor's duties. Responsible for providing own materials and equipment as agreed.

Building operatives
General building personnel, responsible for groundworks, unloading materials and general housekeeping.

▲ Figure 1.1 The building team

Career opportunities

Progression and continuing professional development (CPD) are encouraged in construction. It is not unusual for an operative to start as a construction labourer and then go on to gain experience and qualifications to become a **site manager**.

KEY TERM

Site manager: responsible for the completion of a building project effectively, safely and on time.

2 UNDERSTAND CONSTRUCTION INFORMATION

Types of buildings

As we have seen, the main types of buildings in construction are:

- **residential:** where people live, such as houses, flats and bungalows
- **commercial:** where people shop and purchase items, such as supermarkets, shopping centres and cinemas
- **industrial:** where people go to work, such as factories, warehouses, showrooms and manufacturing plants.

These types of buildings can be categorised as:

1 low-rise buildings, which have only a few storeys and are defined as an enclosed structure below 35 m height which is divided into regular floor levels
2 medium-rise buildings, which are between four and eleven storeys high and are typically office developments, although some are built for residential use
3 high-rise buildings, which are tall as opposed to low and are defined in terms of height depending on any local planning laws. These are mainly hotel and office developments but sometimes residential.

For residential construction, this table shows the different categories of building.

Type of building	Explanation
Detached 	A stand-alone property. It is a free-standing residential construction, referred to as a single family home.
Semi-detached property 	Shares one common wall with the next house. It is a single family home.
Terraced property 	A row of attached dwellings sharing dividing walls. This is medium-density housing, but each property is a single family home.

Building requirements

All buildings must meet minimum requirements under **building regulations** to ensure that developments are safe and healthy. These regulations are enforceable by law. A minimum standard of building work must be completed to adhere to these regulations and materials must be fit for purpose.

Regulations are continually revised and updated and workers in the construction industry need to be aware of new regulations. The most recent update was the Building Regulations 2010 legislation.

Key considerations for a good building:

1 Security: doors and windows should meet safety and security requirements regarding locking and safe opening.
2 Safety: safe exit procedures around fire safety and fire protection.
3 Privacy: considerations for neighbouring properties in terms of acoustic/noise travel and viewing into property.
4 Warmth: up-to-date insulation installed in properties.
5 Light: well-structured windows and openings, letting in natural light.
6 Ventilation: good system to facilitate supply of fresh air, air movement/change, temperature of air, humidity and purity of air.

A well-designed building will meet the minimum standards (or above) of all current building regulations, in line with legal and environmental requirements.

INDUSTRY TIP

You must check if you need approval before you construct or change buildings in any way.

Some works in construction do not need approval from building control, such as:
- most repairs, replacements and maintenance work
- new power and lighting points
- like-for-like replacements of baths, showers, basins and sinks.

Planning permission refers to the approval needed for construction or expansion and sometimes demolition. It is usually given in the form of a building permit and is governed by building control regulations. If planning permission is required, there is a four-stage process:

1 A planning application is submitted to the local council.
2 The planning authority publicises the application, in the press and online.
3 The public has a number of weeks to comment and challenge the application.
4 A decision is made on the application, usually within eight weeks of the application being submitted.

Different parts of a building

A construction has two different parts:

1 The substructure: the part of the building which is constructed below ground level. The substructure passes the **compressive load** of the building to the soil below.
2 The superstructure: the part of the building above ground level where the activities for which the building was designed are usually carried out. This part of the building is the load that is carried by the substructure.

The make-up of the substructure and superstructure are defined by different parts, as shown in this table.

Type of component	Explanation
Primary components	Parts of the building that provide support, floor access, breaking up of space and protection, e.g. foundations, floors, walls, stairs and roofs.
Secondary components	Non-load-bearing parts that are used to close off openings or to provide a finish, e.g. skirting boards, architraves, doors and windows.
Finishing components	Plastering, facing brickwork, rendering and decoration all deemed to be superficial and to add to the décor and finish of a building.
Service components	All electrical, Wi-Fi, plumbing, sewerage and mechanical services.

KEY TERM

Building regulations: rules enforced by the building control departments of local councils to ensure all buildings are safe and fit to live and work in. These regulations contain the minimum standards for design, construction and alterations to buildings.

Compressive load: a weight which tends to shorten or 'squash' a structure.

---

3 BUILDING SUBSTRUCTURE

Buildings vary in types, appearance and the method by which they are built. However, they also have common design features. For example, in the substructure, all buildings have some sort of foundations and the type of foundation chosen will depend on these factors:

- type of building design (timber frame and masonry construction require different foundations)
- type of ground being built on (bedrock, clay and shale require different foundations)
- adequate depth to prevent frost damage
- **bearing capacity** failure
- quality
- adequate strength
- adverse soil changes
- **seismic forces**.

Different types of foundations

Foundations must be able to resist movement in the ground (adverse soil changes), as ground conditions can be very different from one corner of a building to another. When carrying out a **topographic survey**, samples of soil are taken from **bore holes** drilled on the site and the samples sent away for testing. The soil analysis will show:

- the condition of the soil (clay or sandy)
- the depth of the soil
- the depth of the water table
- any contamination of the soil.

This table shows the different types of foundations and whether they are used for **shallow** or **deep foundations**.

Foundation	Shallow	Deep
Strip	✓	
Wide strip		✓
Pad	✓	✓
Pile		✓
Raft	✓	

KEY TERMS

Bearing capacity: the ability of soil to support loads applied to the ground, so as not to produce **shear failure**.

Shear failure: occurs when there is not enough resistance between materials, so structures can move and flex; this leads to structural weakness and cracking.

Seismic forces: forces which act on a building to represent the effect of an earthquake.

Topographic survey: a survey that gathers data about elevation points on a piece of land and presents them as contour lines on the plot. It gives information about the natural and human-made features of the land, such as natural streams or existing groundworks.

Bore hole: a narrow shaft bored (drilled) in the ground vertically or horizontally, to test soil conditions.

IMPROVE YOUR ENGLISH

- **Horizontal** means lying flat (like the horizon), from left to right or right to left.
- **Vertical** means standing or hanging upright, from top to bottom or bottom to top.
- **Parallel** lines never intersect (cross).

It is important you learn what these terms mean, as they are often used in the industry.

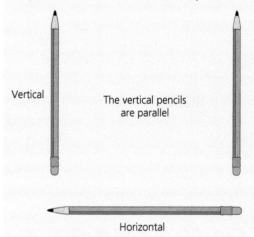

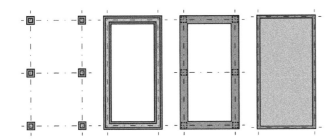

▲ Figure 1.2 Pad, pile, strip and raft foundations

Strip foundation

This type of foundation is used where the soil is of good bearing capacity. It is a shallow foundation used to provide a continuous level (sometimes stepped) strip of support to a linear structure (such as a wall or a closely spaced row of columns) built centrally above the strip. Strip foundations can be filled in two ways:

1 Lay a thin strip of concrete with a minimum depth of 150 mm and then build up to ground-level **damp proof course (DPC)** level with common bricks.

2 Mass fill: fill the majority of the trench with concrete and build to ground-level DPC with trench blocks.

Both types of strip can be strengthened by the addition of steel in the foundation.

KEY TERMS

Shallow foundations: a type of foundation that transfers building loads to the earth very near to the surface.

Deep foundations: a type of foundation that transfers building loads to a subsurface layer or a range of depths.

Damp proof course (DPC): a layer or strip of watertight material placed in a joint of a wall to prevent the passage of water. Fixed at a minimum of 150 mm above finished ground level.

▲ Figure 1.3 Strip foundation trench

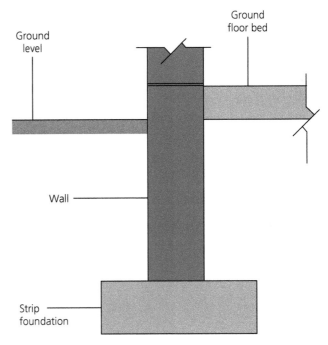

▲ Figure 1.4 Cross-section of strip foundation

Wide strip foundation

This is used where the soil is soft or of a low bearing capacity, to spread the load over a larger area.

This type of foundation is generally considered for deep foundations, as the loads are transferred to deeper layers with a higher bearing capacity. Wide strip foundations are reinforced with steel so the loading per m^3 is reduced.

IMPROVE YOUR MATHS

Concrete is measured in cubic metres, m^3. See page 43 for volume calculations.

What volume of concrete will be required for a floor $43\,m^2$ area with a depth of $65\,mm$?

Pad foundation

This is generally a shallow foundation but can be deep depending on the ground conditions. It is a spread foundation formed by rectangular, square or sometimes circular concrete pads that support single point loads such as structural columns, groups of columns or framed structures. The load is then spread by the pad to the bearing layer of soil or rock below.

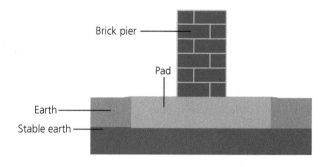

▲ Figure 1.6 Pad foundation

Pile foundation

This is a series of columns constructed or inserted into the ground to transmit loads to a lower level of **subsoil**. A pile is a long cylinder of strong material, such as concrete. Piles transfer the loads from structures to hard strata, rocks or soil with high bearing capacity.

KEY TERM

Subsoil: the layer of soil under the topsoil on the surface of the ground. It is composed of a variable mixture of small particles such as sand, silt and clay, but with a much lower percentage of organic matter and humus (a dark, organic material that forms in soil from plant and animal matter decay).

▲ Figure 1.5 Pad foundation with steel reinforcement

▲ Figure 1.7 Pile foundation

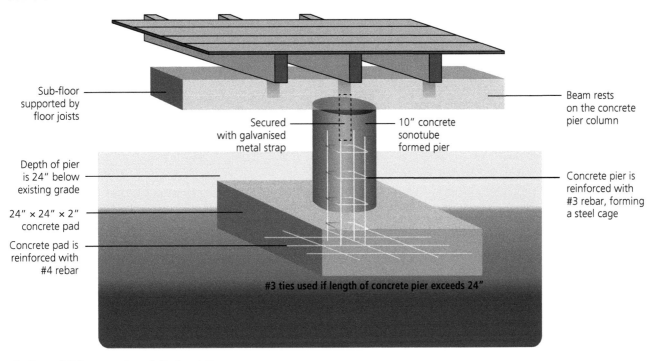

Sub-floor supported by floor joists

Beam rests on the concrete pier column

Secured with galvanised metal strap

10" concrete sonotube formed pier

Depth of pier is 24" below existing grade

Concrete pier is reinforced with #3 rebar, forming a steel cage

24" × 24" × 2" concrete pad

Concrete pad is reinforced with #4 rebar

#3 ties used if length of concrete pier exceeds 24"

▲ Figure 1.8 Cross-section of pile foundation

Raft foundation

This is a thick reinforced continuous slab resting on the soil that extends over the entire footprint of the building. It thereby supports the building and transfers its load to the ground, providing support for several columns and load-bearing walls.

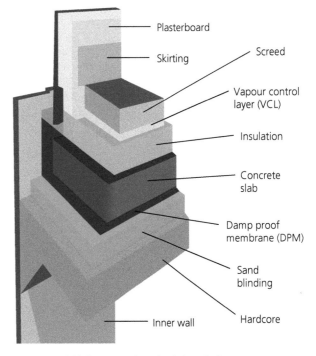

Plasterboard

Skirting

Screed

Vapour control layer (VCL)

Insulation

Concrete slab

Damp proof membrane (DPM)

Sand blinding

Hardcore

Inner wall

▲ Figure 1.9 Raft foundation

▲ Figure 1.10 Cross-section of raft foundation

The table shows the different elements of raft foundation.

Elements of raft foundation	Description
Hardcore	Sub-base of crushed stone, mechanically compacted.
Sand blinding	Provides a clean, level and dry working surface around 50 mm in depth to make sure the hardcore does not puncture the DPM.
Damp proof membrane (DPM)	Polyethylene type material applied to prevent moisture seeping through to the concrete slab.
Concrete slab	Supports a number of columns and load-bearing walls.
Insulation	Protects against energy loss and helps to control moisture permeation.
Vapour control layer	Protects from consequences of condensation: blocks the passage of warm moist air entering the structure. Made of a plastic-type material.
Screed	Flat level floor made with sharp sand and cement to enable choice of floor finish.

Services

When planning foundations, it is often important to consider the type of services which will be installed in a building, as these services have to be incorporated within the foundation structure.

Types of services in buildings include the following:

1 Electricity: ducting is laid in the foundation to allow for electric cable, at a minimum depth of 240 mm. It is identified by black and yellow tape.
2 Gas: this is not allowed under a concrete slab. It is good practice to run a gas pipe around a slab for service. It is identified by yellow sleeving.
3 Water: mains water is facilitated by ducting and brought into the building from the roadside. It is identified by blue sleeving.
4 Drainage: combined system to collect rainwater and domestic sewage in the same pipe and send it to treatment facilities. The pipe can be clay or plastic.
5 Communication networks: cables used to connect and transfer data and information between computers, routers and switches. They are usually identified with purple sleeving/ducting.

4 BUILDING SUPERSTRUCTURE

Once foundations are built, the next phase of the build is the superstructure. This incorporates four main activities, shown in this table.

Type of superstructure	Materials used
Floors	Stone, wood, concrete
Walls	Brick, block, concrete
Roofs	Timber, metal, thatch
Internal finishes	Plaster, decoration

Information on finishes is included in the following sections on floors, walls and roofs.

Floors

In flooring structures, there is always a ground floor: the floor of a building which is level with the ground. Any floors above ground level are numerically ordered, such as first floor, second floor and so on.

There are two types of floor – solid and suspended:

- **Solid flooring** is constructed from hardcore, sand, DPM, concrete, insulation, VCL and screed. Solid floors are substantial structures.
- **Suspended flooring** has a void underneath the structure. The floor can be formed using timber joists, with precast concrete panels using the beam and block system (precast concrete beams are laid and then infilled with concrete blocks) or cast *in situ* with reinforced concrete. The floor structure is supported by external and internal load-bearing walls.

IMPROVE YOUR MATHS

The word 'numerical' relates to numbers. A numerical sentence includes only numbers, such as:

$4 + 5 = 9$

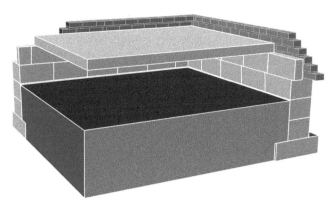

▲ Figure 1.11 Suspended concrete floor

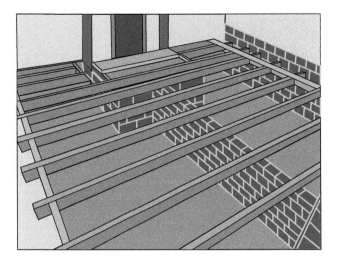

▲ Figure 1.12 Suspended timber floor

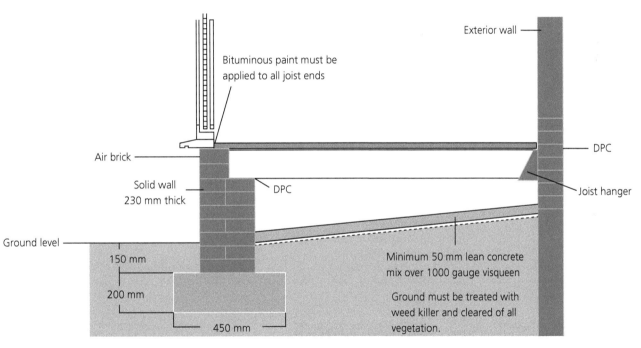

Exterior wall

Bituminous paint must be
applied to all joist ends

DPC

Air brick

Solid wall
230 mm thick

DPC

Joist hanger

Ground level

150 mm

200 mm

450 mm

Minimum 50 mm lean concrete
mix over 1000 gauge visqueen

Ground must be treated with
weed killer and cleared of all
vegetation.

▲ Figure 1.13 Cross-section of suspended floor

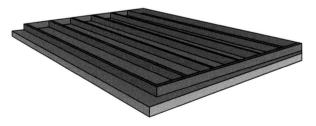

▲ Figure 1.14 *In situ* concrete slab floating floor

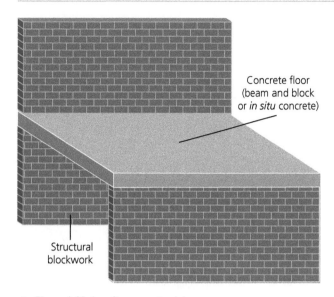

▲ Figure 1.15 *In situ* concrete slab

Some suspended floors are structured from metal and timber, such as the Posi-Joist system. This uses metal web joists which are combined with timber, so the materials are lightweight but have additional strength. This system also allows larger services such as soil

pipes to be fitted into the system. A **sleeper wall** is constructed when a suspended floor is required due to bearing conditions or ground water presence.

KEY TERM

Sleeper wall: a short wall used to support floor joists, beam and block or hollowcore/concrete slabs at ground level.

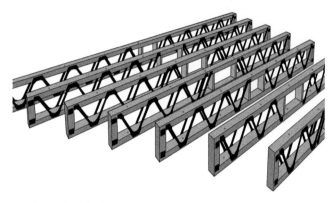

▲ Figure 1.16 Posi-Joists

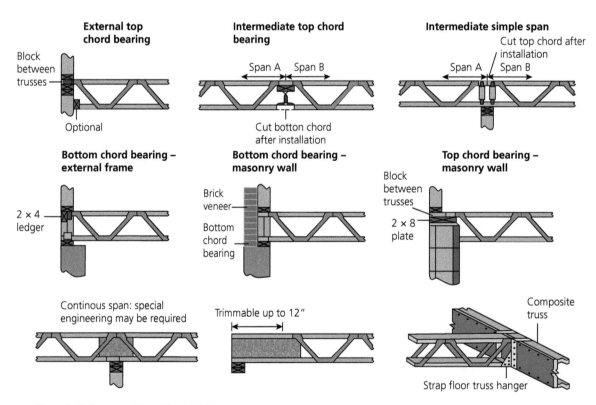

▲ Figure 1.17 Cross-section of Posi-Joists

Also common in suspended floor systems are I-Joists. These are engineered wood joists which have greater strength than conventional wood joists, so they can carry heavier loads and are less likely to bow, crown, twist or split. A disadvantage is that they fail quickly when directly exposed to fire, reducing the time for someone to get out of a building where they have been used.

Timber floors will typically be finished with a variety of coverings such as chipboard or solid timber floor boards.

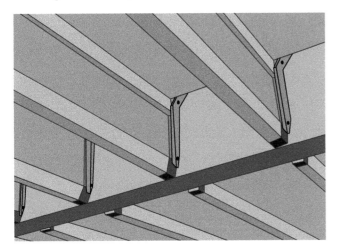

▲ Figure 1.18 I-Joists fixed in position

Upper floors

The style of construction of upper floors has not changed significantly over a long period of time. It is a series of joists supporting a floor covering, with a finish on the underside. This chapter describes the systems used in most buildings, whether in a domestic or commercial build. Depending on the type of floor to be installed, there will be a requirement for beam and block, concrete or timber and screed. The specification will usually require insulation or steel reinforcement, but there will probably be no DPM in upper floors.

Walls

There are two types of wall structure:
- external – forms the external enclosure of a building
- internal – forms walls that divide rooms.

A wall can be a structure that defines an area, carries a load, provides security, shelter or soundproofing and can be decorative. There are many kinds of walls in buildings that form a fundamental part of the superstructure or separate interior rooms.

This table describes the different types of wall.

Type of wall	Description
Load-bearing 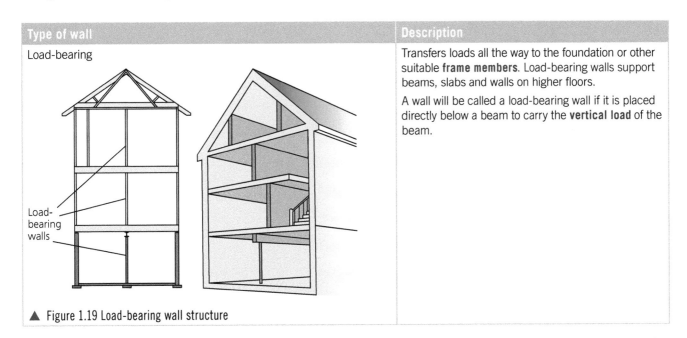 ▲ Figure 1.19 Load-bearing wall structure	Transfers loads all the way to the foundation or other suitable **frame members**. Load-bearing walls support beams, slabs and walls on higher floors. A wall will be called a load-bearing wall if it is placed directly below a beam to carry the **vertical load** of the beam.

Type of wall	Description
Non-load-bearing 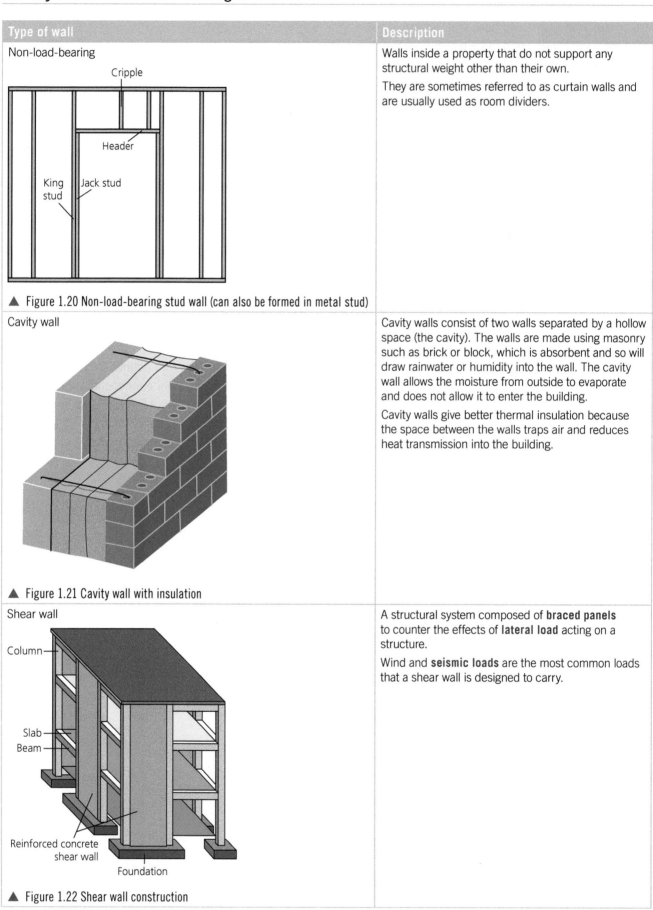 ▲ Figure 1.20 Non-load-bearing stud wall (can also be formed in metal stud)	Walls inside a property that do not support any structural weight other than their own. They are sometimes referred to as curtain walls and are usually used as room dividers.
Cavity wall ▲ Figure 1.21 Cavity wall with insulation	Cavity walls consist of two walls separated by a hollow space (the cavity). The walls are made using masonry such as brick or block, which is absorbent and so will draw rainwater or humidity into the wall. The cavity wall allows the moisture from outside to evaporate and does not allow it to enter the building. Cavity walls give better thermal insulation because the space between the walls traps air and reduces heat transmission into the building.
Shear wall ▲ Figure 1.22 Shear wall construction	A structural system composed of **braced panels** to counter the effects of **lateral load** acting on a structure. Wind and **seismic loads** are the most common loads that a shear wall is designed to carry.

Type of wall	Description
Partition wall 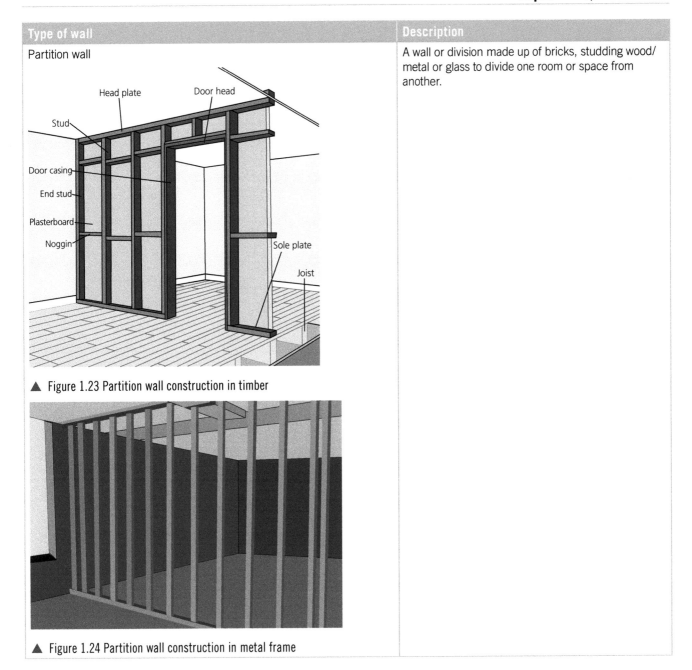 ▲ Figure 1.23 Partition wall construction in timber ▲ Figure 1.24 Partition wall construction in metal frame	A wall or division made up of bricks, studding wood/ metal or glass to divide one room or space from another.

KEY TERMS

Frame members: studs/partitions, wall plates and lintels.

Vertical load: loads in addition to the weight of the structure; this can include the weight of floors, roofs, beams and columns all pushing down compressively.

Braced panels: critical elements of a wood-/metal-framed structure which resist forces that act along the wall plane (mainly to resist lateral wind forces).

Lateral load: typical lateral loads include wind blowing against a facade, an earthquake, or ocean movement on beach-front properties.

Seismic load: relates to forces caused by ground movement such as earthquakes, which will cause movement and possible collapse of structures.

Type of wall	Description
 Panel wall 	Structural system that consists of **planar wall** and slab elements which form an enclosed space. Panels can be made from steel, timber, concrete or masonry.

▲ Figure 1.25 Panel walling in construction

▲ Figure 1.26 Timber frame panel construction

KEY TERM

Planar wall: a flat wall.

Type of wall	Description
Veneered wall 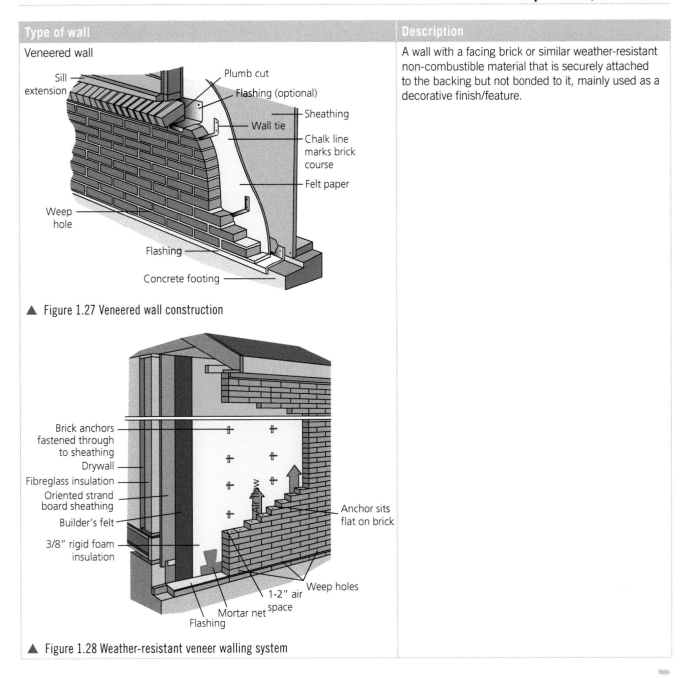	A wall with a facing brick or similar weather-resistant non-combustible material that is securely attached to the backing but not bonded to it, mainly used as a decorative finish/feature.

▲ Figure 1.27 Veneered wall construction

▲ Figure 1.28 Weather-resistant veneer walling system

Type of wall	Description
Faced wall  ▲ Figure 1.29 **Faced** brick wall construction  ▲ Figure 1.30 **Fair-faced** wall construction	A wall where the masonry facing is bonded to the backing material, so the two sections exert common action under load. This means they act together to resist the stresses and strains of any weight above the wall.
Stone wall  ▲ Figure 1.31 Traditional stone house construction	Traditional building material used in construction for many years. Early stone walls were solid, but in more recent times stone is used as an external facing for cavity walls. Dry stone walls are built as a feature for perimeter walls in rural areas.

The main reasons for external walling systems are to:
- protect from weather conditions
- achieve an aesthetic appearance
- fit in well with the existing surroundings.

IMPROVE YOUR ENGLISH

The word 'aesthetic' means to appreciate beauty, which can be created and affected by the order, uniformity, alignment and symmetry of a building.

This table describes some common materials used in walling.

Materials used in walling	Description
Bricks	A building material used to make walls and other elements in construction. A brick is usually a rectangular unit, made from a mixture of clay-bearing soil, sand and lime or concrete materials.
Blocks	Made from cast concrete (**Portland Cement**, aggregate and sand) for **high-density blocks**. **Low-density blocks** might use industrial waste such as fly ash or bottom ash as an aggregate.
Stone	Hard, solid, non-metallic mineral matter (rock is made from stone), which is a traditional and natural building material. A combination of heat and pressure creates blocks of natural stone.
Mortar	Made from a mix of sand, cement/lime and water. It is used to fill the gaps when laying bricks, blocks or other materials used to construct a structure. It sets to a hard consistency.

External wall finishes

Faced and fair-faced finishes will be used for the final décor, while concrete block walls are often given a rendered finish to give either a traditional or modern look (for more details about rendering, see Chapter 4). Faced brick can be made to look aesthetically pleasing by using different types of **bond**.

KEY TERMS

Faced/Fair-faced: these bricks are durable and graded on a scale to match the building material required by the project. Cosmetic face bricks are made to face the world with a smooth look, whereas common bricks/blocks do not have smooth sides.

Portland Cement: also known as Ordinary Portland Cement (OPC), this is the most commonly used cement. It is named after stone quarried on the Isle of Portland off the British coast, as it is similar in colour.

High-density block: durable and resilient, high in strength and with good acoustic rating, generally used for structural purposes.

Low-density block: extremely versatile and can be used in standard wall construction.

Bond [in masonry]: the arrangement of bricks or other building units when building a wall to make sure it is stable and strong. Different types of bond can be used to give a decorative effect.

This table describes some common types of brick bond.

Type of brick bond	Description
Stretcher bond	The long narrow face of the brick is called the stretcher and a stretcher bond is when bricks are laid with the stretcher side showing, the bricks overlapping midway with the bricks in the courses above and below.

Type of brick bond	Description
Header bond 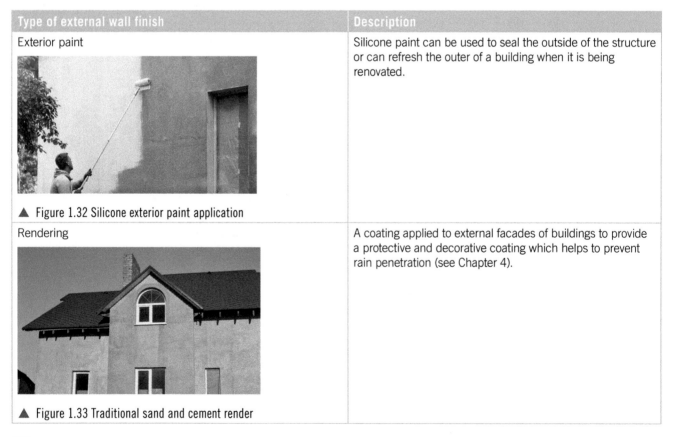	The header is the shorter square face of the brick. A header bond is created by placing the header of the bricks of each course on the faces of walls. The overlap should be half the width of the brick.
English bond	Alternating courses of stretcher and header bonds. Headers are laid centred on the stretchers in the course below. The stretcher rows and header rows should align with each other.
Flemish bond	Headers and stretchers alternate in each course. The next course of brick is laid so that the header is placed in the middle of the stretcher in the course below.

This table describes some types of external wall finish.

Type of external wall finish	Description
Exterior paint ▲ Figure 1.32 Silicone exterior paint application	Silicone paint can be used to seal the outside of the structure or can refresh the outer of a building when it is being renovated.
Rendering ▲ Figure 1.33 Traditional sand and cement render	A coating applied to external facades of buildings to provide a protective and decorative coating which helps to prevent rain penetration (see Chapter 4).

Type of external wall finish	Description
Special coatings  ▲ Figure 1.34 Modern render system, monocouche finish	Modern render systems apply thin coat finish, monocouche finish and thermal cork finish.
External wall insulation (EWI) 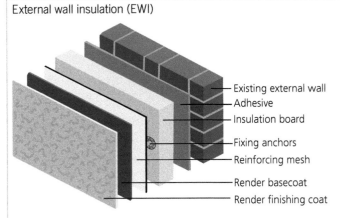 Existing external wall Adhesive Insulation board Fixing anchors Reinforcing mesh Render basecoat Render finishing coat ▲ Figure 1.35 External wall insulation specification  ▲ Figure 1.36 External wall insulation applied to the front of a building	Thermally insulated protection often finished with thin coat render system (see Chapter 4).

Type of external wall finish	Description
Cladding  ▲ Figure 1.37 External wall cladding applied on top of facade	A layer of material that covers another material for protective, thermal and decorative purposes, e.g. insulation, cement carrier boards, timber, metal, brick slips or blended cements.

Modern render systems

- Monocouche finish, which translates to 'one coat' finish, is a scraped texture finish. It is used to give a consistent colour waterproof finish.
- Thin coat finish is a spray- or hand-applied acrylic finish. It is applied over an ordinary render to give a coloured decorative waterproof finish.
- Thermal cork finish is applied by spray. It is an eco-friendly decorative finish applied over ordinary render to give a coloured decorative waterproof finish.

Internal wall finishes

Internal walls are mainly finished with gypsum plasters and decorated with paint/paper finishes and tile finishes in wet areas.

▲ Figure 1.38 Spray cork finish

IMPROVE YOUR ENGLISH
The word 'facade' is used to describe the principal front of a building that faces on to a street or open space.

This table describes some types of internal wall finish.

Type of internal wall finish	Description
Plaster float and skim/set  ▲ Figure 1.39 Applying a backing/floating coat  ▲ Figure 1.40 Applying a skim/set coat	Traditional style of plastering: a backing/undercoat is applied and finished with a skim/set coat for a flat uniform finish, ready for application of décor (see Chapter 3).
Plasterboard direct bond and skim/set  ▲ Figure 1.41 Fixing plasterboard with direct bond	Method of producing a flat wall surface. Sheets of plasterboard are secured to the background surface by application of drywall adhesive to the substrate, leaving air space between board and background. Can be finished for décor purposes by skim/set or tape/joint (see Chapter 3).

Type of internal wall finish	Description
Venetian and Microcement finish  ▲ Figure 1.42 Microcement finishes in wet area	High-end decorative finishes. Trowel applied for feature walls and wet areas.
Emulsion paint finish  ▲ Figure 1.43 Emulsion paint finish	Widely used to decorate walls after new plaster application.
Decorative wallpaper finish  ▲ Figure 1.44 Art deco wallpaper	Decorative finish for walls, sold in rolls. It is applied with glue/paste to decorate after new plaster application.

Roofs

The roof is the structure forming the upper covering of a building. The purpose of any roof is to provide shelter from the elements.

Although there are many different designs of roofs, there are two basic categories: flat and pitched.

Flat roof

This will be almost level but must have a slope of up to 10° to help water run away. Flat roofs are more common in countries with warmer, arid climates, where there is less need for water to run off the roof. Flat roofs can also be used as a living space.

IMPROVE YOUR ENGLISH

The word 'arid' means having little or no rain, too dry or barren.

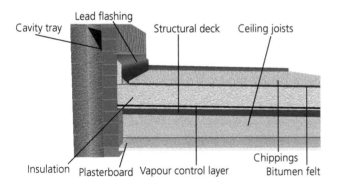

▲ Figure 1.45 Components of a flat roof system

▲ Figure 1.46 Typical modern flat roof

Pitched roof

This is a roof with a sloping surface or surfaces. Its angle is usually more than 20°; some can be more specific, requiring a pitch of more than 10°. The pitch is measured by the vertical rise of the roof divided by its horizontal span and is a measure of its steepness.

There are two types of pitched roof system:

1 Traditional hand cut: each section is individually cut and assembled on site. A traditional cut roof is designed to make sure the load of the roof is evenly transmitted among all of the walls below the roof. The rafters are the main load-bearing elements of the roof.

▲ Figure 1.47 Traditional roof structure

2 Trussed: timber trusses are manufactured off-site and then delivered to be fitted. A truss is a structural framework of timbers which supports the roof. Trusses usually occur at regular intervals and are individual members of the structure, linked by **longitudinal** timbers such as **purlins**. The space between adjacent trusses is known as a bay.

KEY TERMS

Longitudinal: running lengthwise rather than across.

Purlins: roof-framing members that span parallel to the building eaves and support the roofing materials.

▲ Figure 1.48 Trussed roof structure

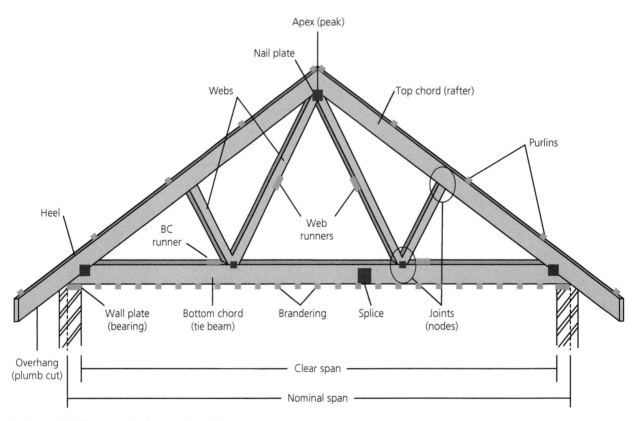

▲ Figure 1.49 Components of an engineered truss

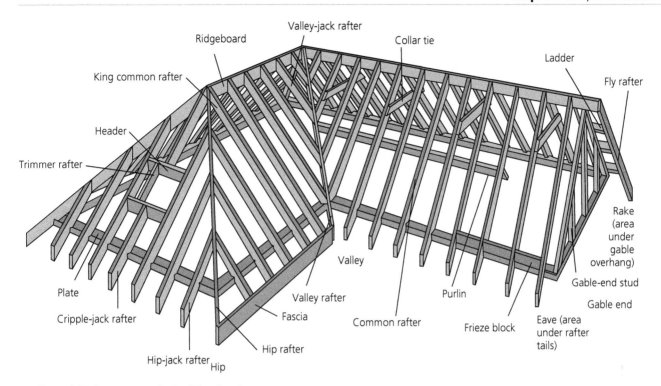

▲ Figure 1.50 Components of a traditional roof

The table shows the many different types of materials used in traditional and trussed roof systems.

Types of material for traditional and trussed roofs	Description
Timber	Cut and manufactured timber from saw mills, using wood from seasoned fir, red cedar and yellow pine.
Lead	Used for flashing coverings and full roof covering, lead has excellent malleability and a low melting point. Can be moulded to any shape and is highly resistant to corrosion and fire.
Slate	This type of rock is used as a roof covering due to its durability, fire-resistance, mould-resistance and low absorption of water. It has an attractive appearance, so might be chosen for decorative purposes.
Tile	Tiles were traditionally made from terracotta or slate, but are now more usually made from durable materials such as concrete and clay.
Felt	Used in waterproof sheeting to cover roofs as an underlay beneath slates or tiles. Felt is an added layer of protection from severe weather. Modern felt is breathable and waterproof.
Sheet	Often corrugated metal sheets made from aluminium, copper, tin, lead.
Synthetic	Steep slope roofing materials such as shingles are synthetic. They are made from different materials such as wood, slate, flagstone, metal, asphalt, plastic and composite materials, or recycled rubber. Asphalt is the most popular as it is easy to install and available in many colours.
Plastic tile	A much lighter type of tile, resistant to ultraviolet radiation. Plastic tiles are more flexible than tiles made of other materials, which allows for complex configuration. Plastic tiles might be used in areas which have extreme weather or temperature changes.
Thatch	Dry vegetation such as straw, water reed, sedge, rushes, heather or palm branches, layered to shed water away from the inner roof. Thatch also acts as a good insulator.

▲ Figure 1.51 Lead flashing on a roof to prevent the passage of water into the building from where the roof joins the structure

ACTIVITY

Find out what the term 'metamorphism' means.

Building elements

Once the shell construction of the superstructure has been built, there are two phases for fitting out the inside of the building.

1 First fix: comprises all work needed *before* putting plaster on the internal walls, including carpentry, plumbing and electrical.

2 Second fix: comprises all the finishing work done *after* the internal walls are plastered, including carpentry, plumbing and electrical.

First fix

This table describes the main first fix elements.

First fix elements	Description
Partitions  ▲ Figure 1.52 First fix timber studding and stairs installation	Stud work to divide the room spaces and accept services. May be made of timber or metal.
Door frames/casings	Linings to receive finished doors.

First fix elements	Description
Windows  ▲ Figure 1.53 Installation of windows  ▲ Figure 1.54 First fix windows with electric cables and back boxes and plasterboard reveal linings	Fitted to seal building openings and to line the reveals, ready for plaster.a
Stairs	Fitted for access to upper floors.
Plumbing  ▲ Figure 1.55 Installation of first fix plumbing	Pipe work for heating, water supply and waste water.

First fix elements	Description
Electrical  ▲ Figure 1.56 Installation of first fix electrics	All electrical wiring and back boxes, not yet connected to mains electricity.
Networks  ▲ Figure 1.57 Installation of network cabling	All network cabling, not yet connected to an external network system.

Second fix

This table describes the main second fix elements.

Second fix elements	Description
Carpentry 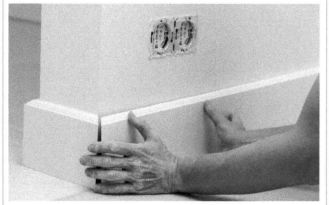 ▲ Figure 1.58 Second fix skirting boards	Fitting of **skirting boards**, **architraves**, finished doors, trims and beading.

Second fix elements	Description
 ▲ Figure 1.59 Second fix internal finished doors	
Kitchens  ▲ Figure 1.60 Installation of fitted kitchen	Fitted kitchen units and worktops.
Bathrooms  ▲ Figure 1.61 Installation of fitted bathroom and sanitary ware and tiling	Baths, showers and tiling.

Second fix elements	Description
Electrical	Connection of cables to mains and fitting of all box connection fronts and light drops.
 ▲ Figure 1.62 Connection of second fix electrics	
Plumbing	Connection of all pipes to mains and sewerage connections. Fitting of showers and all sanitary components and taps.
 ▲ Figure 1.63 Connection of second fix plumbing	
Networks	Connections of all network cabling to mains and supplies.

KEY TERMS

Skirting board: a decorative moulding often made from timber (sometimes plaster) that is fitted at the bottom of a wall to hide the gap between wall and floor and to protect the bottom of a wall from foot traffic.

Architrave: a decorative moulding often made from timber (sometimes plaster) that is fitted around doors and windows to hide the gaps between frames and walls. It also provides a decorative feature.

⑤ INTERPRET BUILDING INFORMATION

Within construction you need to be able to interpret information from many types of document, such as drawings, specifications and schedules. You also need to calculate quantities from drawings and understand drawing scales, symbols and abbreviations.

Information documents used in construction

This table describes the main information documents used in construction.

Type of document	Description
Specifications **INTERNAL AND EXTERNAL PLASTER** a) Siporex walls should generally be plastered on both sides in two coats. Internal walls in industrial buildings may be left unplastered if so desired and joints should be properly finished. b) Block walls should be wetted sparingly before plaster. c) Method of plaster is similar to that of plaster on brick/concrete walls. d) Internal plaster should be carried out in 2 coats with cement mortar 1:3 (one part Portland cement to 3 parts screened and washed sand). First coat may be 4 to 6 mm thickness, plus a second coat (finishing coat). e) External plaster should be carried out in 2 coats with cement mortar 1:3 (one part Portland cement to 3 parts screened and washed sand.) First coat may be 8 to 10 mm thickness and the second coat (finishing coat) may be about 10 mm thick. f) Plastered surfaces should be watered for at least one week. ▲ Figure 1.64 Example of a plaster specification	These are prepared before construction begins and describe how building work or tasks should be carried out by contractors and subcontractors. They provide descriptions that go beyond what photos and videos can explain and contain a high level of detail.
Drawings	These form part of the production information that is incorporated first into tender documentation and then into the contract documents for the construction work. They provide a graphic representation of what is to be built.
Schedule ▲ Figure 1.65 Example of a schedule	A document that lists the works required on a project. The schedule references the requirements of the specifications and contract drawings plus any additional builder's work or fixing extras.

Schedule figure (Figure 1.65):

<Wall Material TakeOff>

A Family and Type	B Material area	C Material:As paint	D Material:Name
Basic wall: interior-4 7/8" partition (1 hr)	800 SF	Yes	Carpet(1)
Basic wall: interior-4 7/8" partition (1 hr)	480 SF	Yes	Carpet(1)
Basic wall: interior-4 7/8" partition (1 hr)	1600 SF	No	Gypsum Wall Board
Basic wall: interior-4 7/8" partition (1 hr)	960 SF	No	Gypsum Wall Board
Basic wall: interior-4 7/8" partition (1 hr)	1600 SF	No	Gypsum Wall Board
Basic wall: interior-4 7/8" partition (1 hr)	960 SF	No	Gypsum Wall Board
Basic wall: interior-4 7/8" partition (1 hr)	800 SF	No	Metal stud layer
Basic wall: interior-4 7/8" partition (1 hr)	480 SF	No	Metal stud layer
Basic wall: interior-4 7/8" partition (1 hr)	800 SF	No	Metal stud layer
Basic wall: interior-4 7/8" partition (1 hr)	480 SF	No	Metal stud layer

Type of document	Description
Bill of quantities *(see table below)* ▲ Figure 1.66 Example of a bill of quantities	A document prepared by the quantity surveyor that provides project-specific quantities of the items of work identified by the drawings and specifications in the tender documentation.
Programme of works *(see Gantt chart below)* ▲ Figure 1.67 A typical programme of works Gantt chart	A reference point for how work will be carried out in specific ways and by specific times.
Building information modelling (BIM)  ▲ Figure 1.68 Example of BIM	An intelligent 3D model-based process that gives architecture, engineering and construction professionals the insight and tools to efficiently plan, design, construct and manage buildings and infrastructure.

Bill of quantities

	A	B	C	D	E	F	G
1	Takeoff Report						
2	Description	Bill Reference	Quantity	Unit	Rate	Markup%	Total
3							
4	**Wall Lining Quants**						
5	WT1 - 1 layer 13 mm plasterboard lining - Nominal 105 mm wall thickness		978.23	m²			
6	WT2 - 2 layer 13 mm plasterboard lining - Nominal 115 mm wall thickness			m²			
7	WT3 - 2 layer 13 mm plasterboard lining - Nominal 130 mm wall thickness		230.74	m²			
8	WT4 - 2 layer 13 mm fire rated plasterboard lining - Nominal 145 mm wall thickness		206.14	m²			
9	WT5 - 2 layer 13 mm fire rated plasterboard lining - Nominal 200 mm wall thickness		278.74	m²			
10	WT6 - 1 layer 13 mm plasterboard lining - Nominal 90 mm wall thickness		36.33	m²			
11	WT6 - 1 layer 13 mm plasterboard lining - Nominal 175 mm wall thickness			m²			
12	S&I fibreglass insulation		865.00	m²			
13	**Total for Wall Lining Quants**						
14	Ceiling Lining Quants						
15	Ceiling type 1 - 13 mm thick perforated flush plasterboard ceiling on furring channel system suspended		566.54	m²			
16	Ceiling type 2 - 13 mm thick flush plasterboard feiling on furring channel system suspended		84.26	m²			
17	Ceiling type 3 - 1200 × 600 mineral fibre ceiling tiles in 2-way exposed aluminium T-bar suspension system		59.06	m²			
18	Ceiling type 4 - 1 layer 50 thick acoustic panels direct fixed to surface of Fyrchek plasterboard; 2 layers 13 thick Fyrchek plasterboard ceiling fixed to underside of ceiling joist		175.60	m²			
19	Ceiling type 5 - 6 thick Versilux lining with 10 mm expressed joints where shown on furring channel suspended.			m²			
20	Ceiling type 6 - new acoustic absorption ceiling tiles in existing grid armstrong 1200 × 600 × 19		911.50	m²			
21	Ceiling type 7 - 1 layer 13 thick perforated flush plasterboard ceiling fixed to underside of ceiling joist; 2 layers 13 thick Fyrchek plasterboard fixed to top of ceiling joists 75 mm R1.5 fibreglass		145.12	m²			
22	Ceiling type EX - Existing mineral fibre tile ceiling replace damaged tiles as required		176.25	m²			
23	**Total for Ceiling Lining Quants**						
24	Cost to removal existing ceiling		176.25	m²			
25	Extra allowance for access		1.00	each			
26							
27	Total						

Type of document	Description

Tender documentation
 	
5. TENDER DOCUMENTS
Tender documents are prepared to seek tenders (offers). Tender documents also help to maintain and protect the relationship between client and contractor.
 	
Document of Tender
1. Invitation for bid
2. Instruction to bidder
3. Standard forms
4. Condition of contract
5. Form of tender
6. Bidding data
7. Contract data
8. Specifications / general and special
9. Bills of quantities (BOQ) with pricing preambles
10. Drawings
11. Schedules of additional information
    • Sample copy of letter of acceptance
    • Sample copy of bank guarantee for performance bond
    • Schedule of day work rates
    • Sample copy of bank guarantee for advance payment
    • Subsoil investigation report
    • Special/particular specifications
Notice to Tenderers
    This document contains a Project Summary, a listing of Tender Documents, key dates, validity period, contact details, number of copies required and details of tender submission location and timing.
Conditions of Tendering
    This document details the overall tender process, including the delivery method, probity issues, communication issues, the criteria for selection and the evaluation process.
	Project- and asset-owners use these documents to invite vendors (contractors, **subcontractors**, suppliers) to bid on projects. By submitting these documents, vendors are volunteering to work on a project.

▲ Figure 1.69 Information likely to be in tender documents

Technical information

Many different types of drawing are used for construction purposes:

1. Working drawings are scale drawings with information on plans, elevations, section details and location of developments. Location drawings are **block plans** and **site plans**, usually giving a bird's eye view of the proposed development with layout of roads, services and drainage.
2. Component range drawings show the range and sizes of components produced by the manufacturer, such as kitchen units.
3. Assembly or detail drawings contain information on how components are put together, such as flat pack shelving.

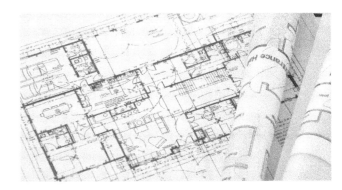

▲ Figure 1.70 A working drawing

All drawings must have a title panel, usually in the bottom left corner of the drawing. This panel will contain the:
- title
- scale used
- name of person who did the drawing
- drawing number/project
- company name
- project title
- date
- revision notes
- projection type.

KEY TERMS

Subcontractor: a tradesperson who is not directly employed by a company but is employed for short periods to complete some aspects of the work. They are paid for the completed work at a set price.

Block plan: drawing that shows the proposed development in relation to the surrounding properties.

Site plan: shows the plot in more detail, with drain runs, road layouts and size and position of existing buildings.

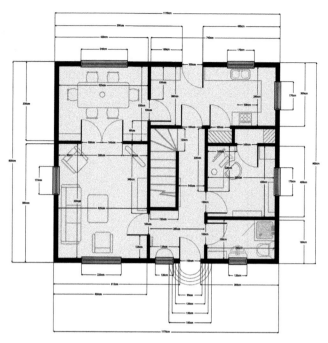

▲ Figure 1.71 This floor plan shows the position of walls, size of rooms and other elements such as staircase position

A section through drawing shows a view of a structure as though it has been sliced in half. This enables the viewer to visualise the whole structure, including the positioning of rooms, windows and doors.

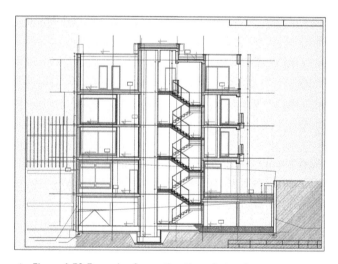

▲ Figure 1.72 Example of a section through drawing

An elevation drawing shows the finished appearance of a given side of a structure. Four elevations are drawn, one for each side.

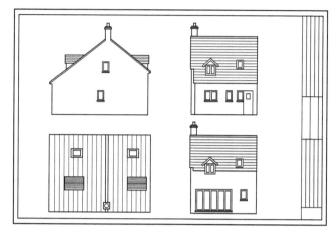

▲ Figure 1.73 Example of an elevation drawing

A detail drawing shows specific details of construction, usually as a cross-section. This provides a detailed description of the geometric form of part of an object. Detail drawings are large-scale drawings that show in great detail parts included in less detail on other drawings. They provide clear, accurate information, such as showing how beads for a render project meet at corner and soffit profiles.

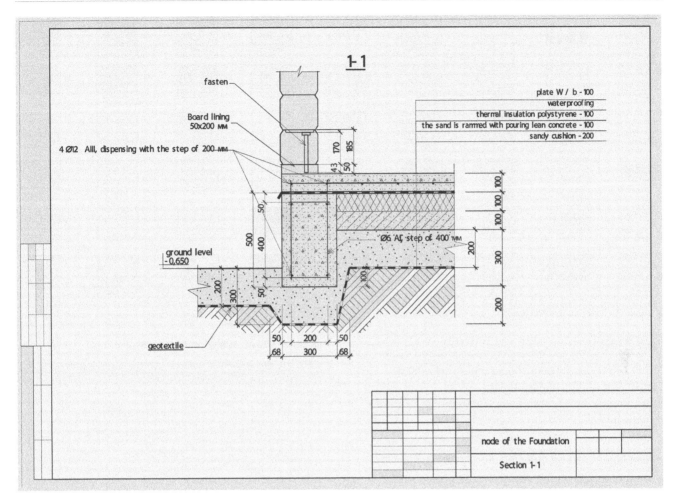

▲ Figure 1.74 Example of a detail drawing

IMPROVE YOUR MATHS

The word 'geometric' is related to geometry. A geometric shape might be a triangle, square or pentagon.

Drawing symbols and hatchings

Hatchings are used as symbols so that different types of material, object and space can be easily identified on a drawing. The hatchings used are standardised so everyone looking at the drawing will know what they stand for. When possible the hatchings should be drawn to scale and they should be consistent for the whole set of drawings. Hatchings take up less space on a drawing than writing all the details in text.

Drawing scales

A drawing that shows a real object with accurate sizes reduced or enlarged is called a scale drawing.

The scale is a ratio of the size of the drawing to the size of the object being drawn. This may be referred to as a **scale ratio**. For example:

- If the scale is 1 cm : 3 cm, then a length of 1 cm in the drawing represents 3 cm in real life (or original size).
- The scale is written as the length in the drawing, then a colon (:), then the matching length on the real thing.
- A scale of 1 cm to 3 cm is written 1 : 3.

The **metric scale** uses the millimetre as its base measurement, so for example 1 : 50 on the metric scale equals one-fiftieth (1/50) of the full size, or 1 millimetre on the drawing equals 50 millimetres in reality.

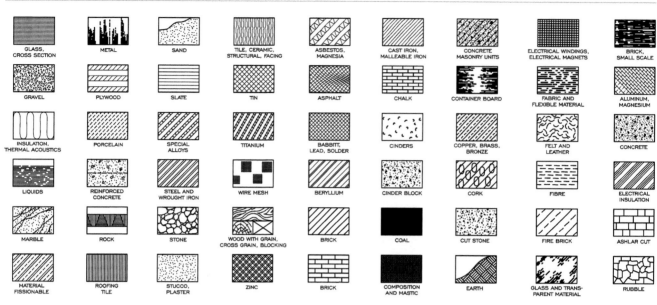

▲ Figure 1.75 Hatchings/symbols

KEY TERMS

Scale ratio: the ratio of the size of a drawing to the size of the object being drawn.

Metric scale: a system of measurement in millimetres.

A small object can be enlarged. For example, if a drawing of an object is twice the size of the object, the scale is 2:1.

A scale ruler is used to determine the actual dimensions of a distance on a scaled drawing. On a metric ruler, the numbers represent centimetres and the individual lines in between represent millimetres. Each millimetre is one-tenth of a centimetre, so ten millimetres equals one centimetre.

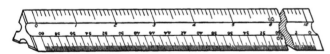

▲ Figure 1.76 A scale ruler

INDUSTRY TIP

When measuring, make sure that one end of the object is lined up with the 0 cm mark on the ruler.

This table shows scales in common use.

Scale	Use
1:1	Full size
1:2, 1:5, 1:10	Building details
1:20, 1:50, 1:100, 1:200	Plans, elevations and sections
1:200, 1:500, 1:1250	Site plans
1:1250, 1:2500	Location plans

IMPROVE YOUR MATHS

5 mm to a scale of 1:10 = 50 mm

$5 \times 10 = 50$

Methods of producing working construction drawings

There are two methods of producing working construction drawings:

1 by hand and drawn to scale by an architect
2 using computer-aided design (CAD) – a computer software system creates precision drawings or technical illustrations to create two-dimensional (2D) and three-dimensional (3D) models.

Common types of drawing produced are:
- orthographic: an object is depicted using parallel lines to project its outline onto a plane; this is a means of representing three-dimensional objects in two dimensions
- isometric: a picture of an object in which all three dimensions are drawn to full scale rather than **foreshortening** them to the true projection; isometric projections are used to visually represent 3D objects in 2D technical drawings.

KEY TERM

Foreshortening: shows an object or view as closer than it is; dramatically reduces an object in scale.

IMPROVE YOUR ENGLISH

The term 'three-dimensional' is used to describe an image which appears to have length, breadth and depth.

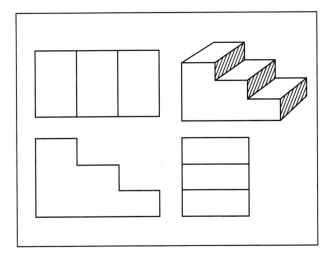

▲ Figure 1.77 Example of an orthographic drawing

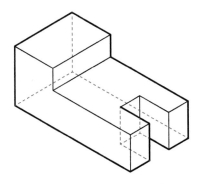

▲ Figure 1.78 Example of an isometric drawing

Calculating quantities when ordering materials

When working in construction, at some stage you will need to calculate:
- quantities of materials needed for projects, such as plasterboard, bags of plaster, floor screed, beads and trims
- the costs of the products
- how many hours/days it will take per person to complete a project.

Without these calculations, a plasterer cannot work out prices and tender for projects. Even if you always work for another contractor, you will need these calculation skills. Plasterers are usually paid per square metre of materials they apply, so you need to be able to calculate your weekly wages to make sure you are being paid correctly.

INDUSTRY TIP

In the apprenticeship training period of your career, you will be paid an agreed amount per week. When fully competent, plasterers are usually paid per square metre of materials applied to surfaces.

It is important to allow 12% extra when ordering materials in bulk to allow for wastage. Otherwise, there may be a loss of profit if extra time and materials are needed to complete a project.

Decimalisation is the conversion of a system of currency and weights and measures to the metric system. This happened in the UK during February 1971.
- Sheet plywood is still measured in imperial units and sold in units of 8 foot × 3 foot.
- Sheet plasterboard is measured in metric units and is sold in units of 2400 mm × 1200 mm.

Points to remember when ordering materials for a project:
1 If possible, buy all the materials for a project at once (this is cheaper than buying in split loads).
2 Ensure there is enough space on site for bulk deliveries (if not, you will have to buy split loads).
3 Make sure you have safe and dry storage to protect materials from damage.
4 Try to order in bulk so that materials are from the same batch to maintain uniformity.
5 Allow for 12% wastage when ordering in bulk.

IMPROVE YOUR ENGLISH

The word 'uniformity' means the state or quality of being uniform, the same. It is important for materials to have a uniform appearance.

The UK construction industry uses metric measurements to do most calculations. However, imperial measurements are occasionally used.

- Metric measurements: units such as metre, centimetre, millimetre, litre and gram are used to measure length, liquid volume and weight.
- Imperial measurements: units such as yard, foot, inch, pint, pound and ounce are used to measure length, liquid volume and weight.

These tables explain the correct units for measurement.

Metric measurement	Units
Length	Millimetre (mm)
Liquid volume	Millilitre (ml)
Weight	Gram (g) (weight is measured in kg or g)

Imperial measurement	Quantities	Example
Length	1000 mm = 1 m	3 mm × 1000 = 3 m
Liquid volume	1000 ml = 1 l	3 ml × 1000 = 3 l
Weight	1000 g = 1 kg	3 g × 1000 = 3 kg

Calculations in standard maths

Addition

The addition of two whole numbers involves finding the total amount. It is shown by the plus symbol (+).

Examples of addition:

$$1 + 2 = 3 \qquad 5 + 4 = 9$$
$$10 + 40 = 50 \qquad 125 + 200 = 325$$

Applied to the job:

An area to be plastered will need 10 bags of Thistle MultiFinish and 7 bags of Thistle Board Finish.

$$10 + 7 = 17 \text{ bags in total}$$

Subtraction

The subtraction of one number from another number involves finding the difference between the numbers. It is shown by the minus symbol (–).

Examples of subtraction:

$$4 - 1 = 3 \qquad 15 - 10 = 5$$
$$60 - 20 = 40 \qquad 170 - 60 = 110$$

Applied to the job:

A wall to be plastered is 22 m². However, there is a window opening in the wall which is 4 m². What is the area to be plastered?

$$22 - 4 = 18 \text{ m}^2$$

Multiplication

For multiplication, a number is added to itself a specified number of times. It is shown by the times symbol (×).

Examples:

$$4 \times 3 = 12 \qquad 10 \times 4 = 40$$
$$25 \times 10 = 250 \qquad 120 \times 160 = 19\,200$$

Applied to the job:

A wall to be plastered measures 4 m by 2 m. To work out the area, we multiply the two numbers together:

$$4 \text{ m} \times 2 \text{ m} = 8 \text{ m}^2$$

Division

We divide or share numbers to find a value. Division is signified by the obelus symbol (÷).

Examples of division:

$$12 \div 3 = 4 \qquad 40 \div 5 = 8$$
$$250 \div 10 = 25 \qquad 1500 \div 300 = 5$$

Applied to the job:

A wall to be plasterboarded is 30 m². The area of one plasterboard is 2.88 m². To work out how many plasterboards are required, we divide the wall area by the plasterboard area:

$$30 \div 2.88 = 10.42$$

Therefore 11 plasterboards will be required.

Working out calculations

- For angle beads or trims, calculations have to be made by linear measurement – this means measuring in a straight line, end to end. Linear length is used when measuring all the sides of a room (perimeter).
- To work out the perimeter of a room, add all the sides together. For example, the perimeter of the room in Figure 1.83 is:

$$3.7 + 2.3 + 3.7 + 2.3 = 12\,m$$

- To work out the area of a square or rectangular room, multiply the length by the width. For example, the area of the room in Figure 1.83 is:

$$3.7 \times 2.3 = 8.51\,m^2$$

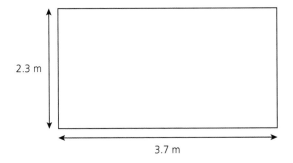

▲ Figure 1.79 Dimensions of a regular room

When a room is an irregular shape, it is easier to:
- split the room into two different sections
- work out each section
- add the sections together to find the area of the irregular shape.

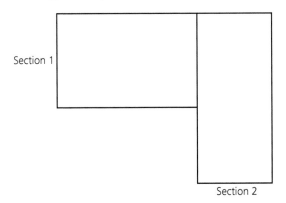

▲ Figure 1.80 Splitting an irregular room

For example, to find the area of the room in Figure 1.84, calculate the area of each rectangular section, then add these two areas together.

An irregular shape including a gable end can be split into two shapes to work out the area. You will have to make calculations for one rectangle and one triangle. To work out the area of a triangle, you have to divide the base by two and multiply by the height.

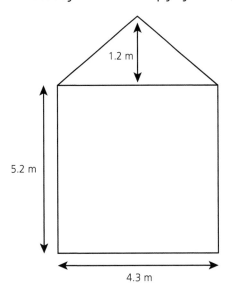

▲ Figure 1.81 Calculating the area of a gable end

IMPROVE YOUR MATHS

Can you work out the area of the gable end in Figure 1.85?

Measuring volume

To calculate volume, we multiply the length by the width by the height.

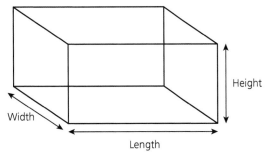

▲ Figure 1.82 Length, width and height of a cuboid

Volume is measured in cube units: mm³, cm³, m³. The volume of the cube in Figure 1.87 is 1 cm³, or:

$$10\,mm \times 10\,mm \times 10\,mm = 1000\,mm^3$$

Remember that:

$$1\,cm^3 = 1000\,mm^3$$

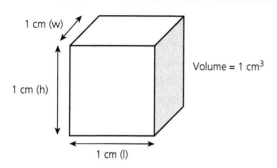

▲ Figure 1.83 Calculating the volume of a cube

Calculating percentages

Area, volume and percentages are important to understand in plastering. As an operative you will have to do various calculations as your development progresses. To calculate for a floor screeding project, you will have to work out the area of floor to be screeded and the amount of material required for the project. When calculating costs, you will need to use percentages to calculate wastage and to work out the value added tax (VAT) on the cost of materials.

There will be VAT on all materials purchased; in 2020, the rate is 20%. One way to work out a percentage is to divide the number by 100 and then multiply it by the percentage required. For example: £200 materials plus VAT @ 20%:

$200 \div 100 = 2$

$2 \times 20 = 40$

VAT on materials $= £40$

Total cost of materials including VAT $= £240$

INDUSTRY TIP

Value added tax (VAT) is a type of consumption tax that is applied to most purchases of goods or services.

IMPROVE YOUR MATHS

Work out VAT on materials costing:

1 £50
2 £80
3 £140
4 £180.

ACTIVITY

A floor has a surface area of 22 m². The agreed price for screeding is £18.75 per m² for labour only. The payment for this project will be:

$18.75 \times 22 = £412.50$

To calculate materials for this floor screed:

The floor area is 22 m² and the screed will be applied to a depth of 65 mm. We cannot multiply metres by millimetres so we write 65 mm as 0.065 m (move the decimal point two places to the left).

The volume can then be calculated:

$22 \text{ m}^2 \times 0.065 \text{ m} = 1.43 \text{ m}^3$

You need to order 1.5 m³ of screed to complete the floor.

Screed costs £16 per m³. Calculate the total cost of the screed, including VAT at 20%.

⑥ SETTING UP AND SECURING WORK AREAS

Planning the site layout

When setting up a construction site, there are five key points to consider:

1 Create a secure perimeter fence to limit access and protect the public.
2 Establish a workflow in line with health and safety rules.
3 Have all paperwork and documentation in order.
4 Get equipment needed for the work and staff.
5 Sort out necessary storage, office space and welfare facilities.

The purpose of planning the infrastructure of site facilities is to make sure that the positions of, and routes between, temporary site structures (such as storage cabins, welfare facilities and site offices) are carefully planned so as not to impact the actual build.

A well-planned site layout should provide safe and clean conditions for working. All site access and egress and pedestrian and vehicular routes should be clearly signposted and kept clear at all times, so that the construction work is not obstructed. If the layout is not planned effectively at the beginning of the project, it can be expensive to reorganise facilities after the build has begun.

IMPROVE YOUR ENGLISH

The word 'egress' means the action of leaving a place.

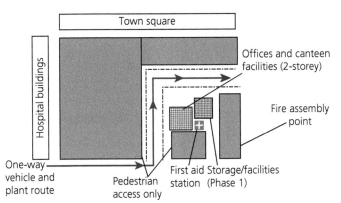

▲ Figure 1.84 Typical entry into a construction site compound

▲ Figure 1.85 Typical site layout

This table describes and explains the essential elements of the site layout.

Planning site layout	Description
Material deliveries	Deliveries should: ● be checked in at the access gate ● follow the vehicular route ● be safely unloaded, either manually or with mechanical handling techniques ● be signed for.
Material storage	Adequate storage facilities and space are required to keep materials safe and dry.

Planning site layout	Description
Neighbouring properties	It is important to ensure residents or users of nearby properties will not be disturbed, in terms of noise, people talking (e.g. loud or offensive language) and start and finish times.
Parking	Safe parking on site is required, with a reverse parking policy in place. Consider nearby properties and any parking restrictions near the site.
Waste management and recycling	There must be an effective waste management programme, with facilities for recycling, safe disposal of contaminated waste and separate skips for different types of waste.
Protection of the natural environment	Protect any natural habitat such as protected wildlife or trees with preservation orders in place.
Access/ egress	Safe access/egress to and from the site for vehicles and pedestrians is essential
Plant	The safe movement and storage of plant is necessary.

▲ Figure 1.86 Typical construction plant

ACTIVITY

Use a suitable search engine to find out what a tree preservation order (TPO) is.

KEY TERM

Plant: machinery, equipment and apparatus used for an industrial activity. In construction, plant refers to heavy machinery and equipment used during construction works such as diggers, dumpers and cranes.

Welfare considerations

A well-planned site will observe the welfare considerations explained in this table.

Welfare considerations	Description
Toilets	Male and female and disabled facilities
Washing facilities	• Hot and cold water washing facilities • Hand sanitiser • Three-stage skin protection station
Storage of personal items	Clothes/coat pegs, lockers
Canteen area	Clean eating and drinking area
Drying room	Room for drying clothes that become wet during works operations

▲ Figure 1.87 Three-stage skin protection station before commencement of work, after completion of work activities and to restore and moisturise skin after washing

▲ Figure 1.88 Typical drying room with personal storage lockers and clothes/coat pegs

Welfare provision on any size of construction site is fundamental in safeguarding the health and well-being of workers. The provision of toilets, a supply of hot and cold water for washing, changing facilities, drinking water and somewhere to eat and rest are basic expectations in modern construction.

It is important when setting up sites to consider the public, employees, materials and tools and equipment. The purpose of a perimeter fence around a construction site is to protect the public and to provide security for the site. Essential public access must be by appointment and visitors must complete an induction process and then sign in and out when attending. This system also applies to employees: all employees should complete an induction when they first start working on site and they should sign in and out whenever they enter or leave the site.

There must be secure storage for materials, tools and equipment. Locked cabins are usually used for this purpose.

ACTIVITY

Use a suitable search engine to find a construction company's induction process.

HEALTH AND SAFETY

It is very important to be safe when working in construction. When you first set foot on a construction site, you will have a health and safety induction before you are allowed to carry out any type of work. A good health and safety induction will cover these points:

1 Health and Safety at Work Act (HASAWA)
2 Reporting of Injuries, Diseases and Dangerous Occurrences Regulations (RIDDOR)
3 Control of Substances Hazardous to Health (COSHH) Regulations
4 Construction (Design and Management) (CDM) Regulations
5 Provision and Use of Work Equipment Regulations (PUWER)
6 Manual Handling Operations (MHO)
7 Personal Protective Equipment (PPE)
8 working at height
9 control of noise
10 welfare facilities
11 toolbox talks.

You can read more about health and safety requirements and legislation in Chapter 2.

7 COMMUNICATION

Communication within all industry is very important to achieve work of high standards, on time and in budget. Put simply, communication means passing on information from person to person. Employees at all levels need good communication skills: without these skills and clear communication lines, businesses cannot function properly. The essence of communication is the ability to share thoughts, information and ideas in a clear way that will not be misunderstood.

Positive and negative communication

Communication can be positive or negative.

Positive communication

Positive communication elicits good responses and actions from the intended audience. It has the power to convert negative feelings into positive ones and helps create an upbeat atmosphere for everyone. Good practice for positive communication is to:

- be brief
- be specific
- be positive
- offer to help.

Negative communication

This is a style of communication in which information is communicated in an unnecessarily negative or harsh way. It is bad practice:

- to lead with the problem
- not to let go of mistakes
- to deal with another person angrily.

Written communication

There are many different kinds of written communication, such as:

- electronic (email, SMS): sent via computers or phone networks
- memorandum: a type of letter containing a statement, usually written by management for the purposes of sharing information
- letter: typed or handwritten, sent in an envelope by post, considered formal and private

- notes: informal short messages, usually sent as a reminder
- drawings: important documentation regarding building works.

There is physical proof of written communication and these messages can easily be shared and reread to gain an understanding. However, some forms of written communication can take longer to reach their destination, can be lost and can be misunderstood if the writing is not legible.

IMPROVE YOUR ENGLISH
The word 'legible' means clear enough to read.

Verbal communication

Verbal communication means using words and speech to share information, such as:

- advising others on a course of action
- expressing agreement
- offering feedback constructively
- giving praise
- reasoning and countering in conversation.

This is the most common method of communication and can be face to face or electronic (via video or walkie-talkie). Communication is instant and misunderstandings can be quickly explained. However, unless it is recorded, verbal communication can easily be forgotten or changed when passed on to other people.

Body language

Body language is a physical form of communication. We communicate attitudes and feelings by our conscious and unconscious movements and postures. These include:

- a shake of the head
- different forms of facial expression such as a frown or smile
- hand movements/gestures
- whole/partial body posture.

Body language gestures can be used effectively to eliminate misunderstandings. However, these gestures can be less effective if there is a distance between the people trying to communicate.

The table shows examples of good and bad body language during conversation.

Examples of good body language	Examples of poor body language
Open body position (arms uncrossed)	Rolling your eyes
Upright posture	Yawning
Relaxed and open facial expression	Hands in pockets
Arms relaxed by your sides	Crossed arms
Positive eye contact	Frowning

Communicating with other people

On a construction site, you might need to communicate with many different people, such as:

- site managers
- supervisors
- other operatives
- members of the public
- architects
- local residents.

You might also occasionally need to communicate with professional bodies, such as:

- the HSE
- local authorities
- planning and building control
- CDM representatives
- environmental agencies
- other trade managers.

It is important to have good, positive communication skills when engaging with all stakeholders, to maintain a professional persona.

Site paperwork

A huge part of communication on site is written documentation, which comes in many forms. It is essential to the successful running of a construction site and involves communication between all stakeholders.

This table lists different types of written document for construction. They are often in digital form.

Type of paperwork	Description
Timesheet	A method for recording the amount of time a worker has spent on each job.

Weekly Time Sheet

NAME OF EMPLOYEE						FOR WEEK ENDING		
DEPARTMENT						EXEMPTIONS		

DAY OF WEEK	MORNING		AFTERNOON		OVERTIME		FOR OFFICE USE ONLY	
	IN	OUT	IN	OUT	IN	OUT	REGULAR HOURS	OVERTIME HOURS
MONDAY								
TUESDAY								
WEDNESDAY								
THURSDAY								
FRIDAY								
SATURDAY								
SUNDAY								
TOTAL HOURS								

NO PERSON PERMITTED TO WORK OVERTIME WITHOUT SPECIAL AUTHORISATION
THIS TIME SHEET MUST BE PERSONALLY FILLED OUT AND SIGNED BY EMPLOYEE.

AUTHORISATION OF OVERTIME ———————————— EMPLOYEE SIGNATURE ————————————

Type of paperwork	Description
Job sheet	This contains instructions to help a worker do their job. It includes details such as the time it takes to perform tasks, the materials required and address information.

CDS
JOB / SERVICE SHEET

Date:	Call/Ref:	Expected time
	JOB NO.	

Person attending

Customer name:	Tel. no.:
Address:	Work:
	Mobile:
	Other:
Postcode:	Who gave you the call:

Work to be carried out:

Work carried out including materials:

Further action required:

Customer signature:	CDS signature:

Type of paperwork	Description
Variation order	This is issued whenever there is a change to contracted work, such as: <ul><li>adding or omitting work</li><li>increasing or decreasing quantity of work</li><li>changing the character or quality of materials</li><li>changing the order of work.</li></ul>
Requisition order	A document generated by a particular section of the company which tells the purchasing department which items need to be ordered.

Variation order

YOUR COMPANY DETAILS

Variation Order

Name:
Address:

Date:
Customer order no.:

Works		Materials	Time	Price

Date	Signature	Print		
			Sub total	
			VAT	
			Total due	

Requisition order

CONSTRUCTION CHANGE ORDER REQUEST FORM

Change Order No.: _____   Contract No.: _____
To: _____   Date: _____
_____
_____

Project Name: _____
Under our AGREEMENT dated _____ _____ (Year)
****
You are hereby authorised and directed to make the following change(s) in accordance with the terms and conditions of the Agreement:

(DESCRIPTION OF THE CHANGE)

FOR THE Additive (Deductive) sum of: _____ (£ _____).

Original Agreement Amount   £ _____
Sum of Previous Changes   £ _____
This Change Order Add (Deduct)   £ _____
Present Agreement Amount   £ _____

The time for completion shall be (increased/decreased) by _____ (___) calendar days due to this Change Order, accordingly, the Contract Time is now _____ (___) calendar days and the substantial completion date is _____. Your acceptance of this Change Order shall constitute a modification to our Agreement and will be performed subject to all the same terms and conditions in our Agreement indicated above, as fully as if the same were repeated in this acceptance.

The adjustment, if any, to this Agreement shall constitute a full and final settlement of any and all claims arising out of or related to the change set forth herein, including claims for impact and delay costs.

The Contract Administrator has directed the Contractor to increase the penal sum of the existing Performance and Payment Bonds or to obtain additional bonds on the basis of a £25,000.00 or greater value change order. 9 Check if applicable and provide written confirmation from the bonding company/agent (attorney-in-fact) that the amount of the Performance and Payment Bonds have been adjusted to 100% of the new contract amount.

Accepted _____ _____ (Year)
By: _____   By: _____
          Contractor   Architect/Engineer
By: _____
          Owner

F0030 (Revised 03/09/19)

Type of paperwork	Description
Delivery note	A document included when goods are sent out to be delivered. It contains a description of the goods and amounts enclosed, but it does not include costing of items. The delivery note must be signed by the recipient.

Delivery Note

Plasterers Ltd
Terrace Rd
Manchester
Tel: 0161 000000

Customer Invoice Address	Customer Delivery Address
Builders Ltd Carmelite House Newcastle Tel: 0191 000000	North Rd Construction Sheffield Tel: 0181 000000

Despatch Date:	5th May 2003
Customer Order No.:	CB 23510
Account No.:	CB 222468

Cat. No.	Quantity Order	Description	Quantity Delivered	Quantity to Follow
232	15	Bags of Multi Finish 25 kg each	15	
234		1200 mm × 2400 mm ×12.5 mm plasterboard	20	

Received in Good Condition:

Signature................................. **Print Name**.........................

Delivery record	A document, usually issued monthly, to track what customers have purchased over a period of time.

Invoice	A document issued by a seller to the purchaser that shows quantities and costs of products purchased.

INVOICE

Contractor/Freelancer		Client	
Name		Name	
Address		Address	
City	Postal code	City	Postal code
Email		Email	
Telephone (Business)	Fax	Telephone (Business)	Fax
Type of contracting			

Changes

Description of Work Performed	Duration of Work		Hourly fee	No.of hours	Amount
	From	To			
					00.00
					00.00
					00.00
					00.00
					00.00
					00.00
					00.00
					00.00
					00.00
					00.00
				Total (before tax)	00.00
	VAT Registration number			VAT	00.00
				Total	00.00

Fee Schedule

If the fee is going to be paid during several months

☐ Jan ☐ Feb ☐ Mar ☐ Apr ☐ May ☐ Jun ☐ Jul ☐ Aug ☐ Sep ☐ Oct ☐ Nov ☐ Dec

Description of fee schedule

Confirmation

Place and Date	Place and Date
Signature, Contractor / Freelancer	Signature, Employer
Print Name	Print Name

Risk assessment	See Chapter 2
Method statement	See Chapter 2

⑧ SUSTAINABILITY

Sustainability means carrying out activities without depleting the environment's resources or having a harmful impact. It is good practice in the construction industry to be sustainable wherever you are working. The principle of sustainability aims to meet the needs of the present without reducing the ability of future generations to meet their needs. In construction terms, this means creating structures and using processes that are friendly to the environment, responsibly sourced and energy-efficient for a building's lifespan.

Fossil fuels such as coal, gas and oil are taken from the ground, but there is a finite amount so they are classed as unsustainable fuels. However, there are infinite natural power sources such as wind, sun and tides, which are classed as sustainable.

IMPROVE YOUR ENGLISH

The word 'infinite' means endless; 'finite' means there is an end or limit to something.

One particular cause for concern within construction is the worldwide use of timber. Some countries have done untold damage to the planet and environment for a number of years by harvesting massive amounts of trees for construction purposes. Taking timber from the world's rainforests has caused many problems to the environment, destroying plants and animals, with some species now lost to the world forever. Trees not only produce the oxygen we need to breathe but they also absorb a lot of the damaging carbon dioxide produced by fossil fuels.

The cutting down of trees also has an effect on the world's climate and water cycle. Practitioners in many industries, including construction, are starting to source more sustainable materials in the hope of regenerating forests and the environment for future generations. Organisations such as the **Forest Stewardship Council (FSC)** have been set up to manage forests sustainably.

KEY TERM

Forest Stewardship Council (FSC): an international, non-governmental organisation dedicated to promoting the responsible management of the world's forests.

Sustainable materials

Good practice in construction includes:
- sourcing materials local to the site to save on transport costs and fuel use
- sourcing timber from properly managed forests which are replenished regularly
- carefully selecting building materials, making direct comparisons between timber frame, metal frame or masonry construction to achieve the most sustainable outcome.

This table describes some sustainable materials used in construction.

Sustainable materials used in construction	Description
Bamboo 	Natural composite material with a high strength to weight ratio; very good structurally.

Sustainable materials used in construction	Description
Straw bales 	Wheat, rice, rye and oats are used to make straw bales as structural elements; good building insulation.
Recycled plastic 	Recycling helps to save energy and landfill space. Recycled plastic is frequently used in construction applications, e.g. roofing tiles, insulation, PVC windows, fences and floor tiles. It can be mixed with virgin plastic to reduce costs without loss of performance.
Wood	If foresting is managed properly, wood is sustainable for future generations.
Rammed earth 	A technique for foundations, flooring and walls using natural raw materials such as earth, chalk, lime or gravel which have been compressed for strength.

Sustainable materials used in construction	Description
Ferrock 	Blocks made from small particles of steel mixed with silica from crushed glass. The iron content corrodes when exposed to air and forms iron carbonate. Ferrock means ferrous rock and carbon is the key element in the glass.
Timbercrete 	A blend of sawmill waste, cement, sand, binders and non-toxic deflocculating additive, which is cured using the renewable resources of sun and wind into a unique building block.
Cobs 	Natural building material made from subsoil, water, fibrous organic material (straw) and lime. It can be modified with sand and clay.

KEY TERM

Deflocculating additive: a substance added to a mixture, to give a slurry that would otherwise be very thick and gooey a thin, pourable consistency.

Sustainable materials used in construction	Description
Lime 	This is a substance containing calcium and composed of oxides and hydroxide, usually calcium oxide or calcium hydroxide. It is breathable material made from ground limestone rock.
Sheep's wool 	100% pure wool can be used as an insulator and is extremely friendly to the environment. This is a sustainable and renewable resource which helps reduce the carbon footprint of a building.

Energy performance

Energy Performance Certificates (EPCs) set out the energy efficiency rating of buildings. This is now a requirement when buildings are built or upgraded, sold or rented. Energy is expensive and its production can be damaging to the atmosphere and environmental surroundings.

Energy-efficient buildings are designed to significantly reduce the amount of energy needed for heating and cooling, regardless of the energy and equipment chosen to heat or cool the building. If systems can be put in place that allow a building to produce its own energy, this is considered good practice.

ACTIVITY

Using a suitable search engine, find out as much information as you can about EPCs.

Renewable energy resources

While some renewable energy systems can be expensive to set up, over many years these systems are cost-effective and sustainable. This is in contrast to fossil fuels such as coal, petroleum and natural gas, which have been our main sources of energy for many years and have proven to be hugely expensive and damaging.

This table describes the main renewable energy resources.

Renewable energy resources in construction	Description
Solar energy 	Radiant light and heat from the sun are harnessed using a range of technologies such as solar heating.
Geothermal energy 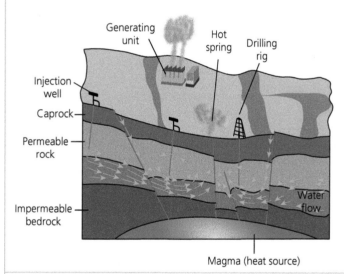	Energy is generated and stored in the ground. The Earth's internal heat is called thermal energy. Geothermal energy originates from the formation of the planet and from radioactive decay of materials. This can be drilled into and used as an energy source.
Heat pumps/ground energy 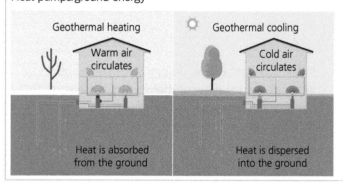	Heat pumps absorb energy from the sun warming the ground. Pipes buried underground extract solar energy and convert this energy into heat.

Renewable energy resources in construction	Description
Tidal power energy 	A form of hydropower that converts energy obtained from tidal movement of the sea into power for electricity.
Wave power energy 	This system converts the up and down movement of waves into electricity. Equipment on the surface of oceans captures energy produced by the movement of the waves.
Hydroelectric energy 	This system harnesses the power of water in motion, such as water flowing over a waterfall, to generate electricity.

Renewable energy resources in construction	Description
Wind energy 	A process by which wind is used to generate electricity, e.g. wind turbines convert kinetic energy in the wind into mechanical power.
Biomass energy 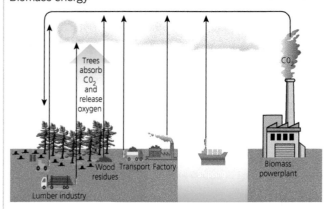	Plant or animal material is used for energy production or in various industrial processes as raw material for a range of products. This system uses purposely grown energy crops such as wood or forest residue, waste from food crops, horticulture, food processing, animal farming or human waste from sewage plants. ● Biomass is burned in a furnace to generate hot gases which are fed into a boiler to generate steam. ● The steam is expanded through a steam turbine or engine to produce mechanical or electrical energy.

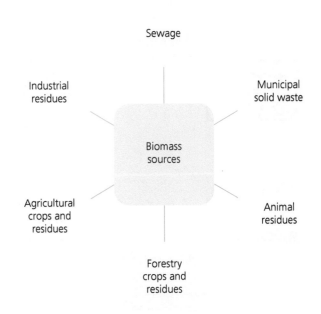

Sewage

Industrial residues

Municipal solid waste

Biomass sources

Agricultural crops and residues

Animal residues

Forestry crops and residues

▲ Figure 1.89 Sources of biomass energy

Protecting the environment

As well as using sustainable energy within construction, it is important to minimise building waste. Construction by its nature produces large amounts of waste materials and legislation now ensures that this waste is managed in an environmentally friendly way.

Building contractors try to get the best possible use from supplies and materials and choose products and methods that reduce waste. There are many recycling processes in construction and building waste is now commonly segregated into different skips to enable maximum recycling of waste materials.

It is important to protect water, plant life and animal species in areas undergoing construction projects, as there is government legislation in place to protect the biodiversity of local ecologies. Construction companies should aim to complete projects on time and to a high standard with the smallest possible impact on the environment.

ACTIVITY

Use a suitable search engine to find out what the **Environment Agency (EA)** is responsible for.

CASE STUDY

Passive house build

Gemma is a project manager for a building company and has been tasked with managing the build for a detached property in a rural area. This will be the first project the company has completed as a passive house build. Gemma wants to show her line manager and future stakeholders that she is capable of managing the project from start to finish.

Her first role is to gather information on **passive housing** projects and speak to designers and architects who have experience in this field. When she has enough information, Gemma draws up a programme of works for the build. The site is fairly rural and there are some weather considerations – these two factors might affect the build in various ways, including sourcing and delivery of materials. Gemma researches the long-term weather forecast and finds material suppliers who are closest to the development. She also researches and engages specialist contractors for some of the technical elements of the passive build: this is the company's first passive build, so they do not have existing contractors with the skills required. Working with new contractors could present quality issues.

Once Gemma has all the quotes for work and materials information, she can instruct contractors and manage the build. If she has planned carefully and allowed for any contingency, all her hard work should pay off and she will be able to impress her employers and all stakeholders.

KEY TERMS

Environment Agency (EA): public body working to protect and improve the environment.

Passive housing: creating an ultra-low energy building with a small ecological footprint that requires little energy for space heating or cooling.

Test your knowledge

1 Which one of these would you not see on a specification for strip foundations?

   A the concrete mass

   B the location of reinforcing bars

   C the width of the strip

   D the depth of the hardcore bed

2 Which of these can be taken from a block plan and a site plan?

   A services location

   B room layout

   C north direction

   D datum

3 Which of these is not involved in the tender process?

   A closed bid

   B deposit payment

   C time of opening bids

   D date for return

4 Where would you get information on foundations, walls, materials, surface finishes, floors, roofs and components related to a building project?

   A specification

   B bill of quantities

   C tender document

   D drawing

5 What is a possible implication and repercussion of inaccurate estimating?

   A Projects will be profitable.

   B There will not be enough labour to do the work.

   C The company will not make profits.

   D All contracts will be won.

6 How can good communication improve teamwork?

   A It creates argument and conflict.

   B Workers will have poor organisational skills.

   C It gives opportunities for training.

   D It improves morale in the workplace.

7 What is the hatching symbol below?

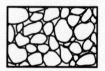

   A blockwork

   B timber

   C stone

   D gravel

8 Which brick bond is shown below?

   A English

   B stretcher

   C Flemish

   D header

9 Which scale should be used for a site plan?

   A 1:1

   B 1:5

   C 1:200

   D 1:2500

10 What is being created in this image?

   A biomass energy

   B tidal power energy

   C hydroelectric energy

   D wave power energy

INTRODUCTION

The aim of learning about health and safety in the workplace is to understand the essential tasks for achieving a healthy and safe workplace. This chapter will help you to control and identify hazards and risks and show you how to organise, plan, monitor and review health and safety in construction. This information is relevant for anyone who is involved in construction work, including designers, contractors, operatives, clients and architects.

By reading this chapter you will gain knowledge of:

1 health and safety roles and responsibilities: the Health and Safety Executive (HSE)
2 health and safety legislation:
   - Health and Safety at Work Act 1974
   - Control of Substances Hazardous to Health (COSHH) Regulations
   - Provision and Use of Work Equipment Regulations (PUWER)
   - Control of Noise at Work Regulations
   - Reporting of Injuries, Diseases and Dangerous Occurrences Regulations (RIDDOR)
   - Working at Height Regulations
   - Electricity at Work Regulations
   - Personal Protective Equipment (PPE) at Work Regulations
   - Manual Handling Operations Regulations
   - Construction (Design and Management) Regulations
   - Control of Vibration at Work Regulations 2005
3 construction site health and safety inductions
4 accidents and emergencies
5 fire safety
6 safety signs and signals
7 welfare at work
8 first aid.

The table below shows how the main headings in this chapter cover the learning outcomes for each qualification specification.

Chapter section	Level 1 Diploma in Plastering (6708-13) Unit 201/601	Level 2 Diploma in Plastering (6708-23) Unit 201/601	Level 2 Technical Certificate in Plastering (7908-20) Unit 201
Health and Safety Executive	Learning outcomes 1.1, 1.2, 1.3, 1.4	Learning outcomes 1.1, 1.2, 1.3, 1.4	Topic 2.4
Control of Substances Hazardous to Health (COSHH) Regulations	Learning outcome 1.1	Learning outcome 1.1	Topic 2.4
Provision and Use of Work Equipment Regulations (PUWER)	Learning outcome 1.1	Learning outcome 1.1	Topic 2.4

Chapter section	Level 1 Diploma in Plastering (6708-13) Unit 201/601	Level 2 Diploma in Plastering (6708-23) Unit 201/601	Level 2 Technical Certificate in Plastering (7908-20) Unit 201
Control of Noise at Work Regulations	Learning outcomes 1.1, 4.2	Learning outcomes 1.1, 4.2	Topic 2.4
Reporting of Injuries, Diseases and Dangerous Occurrences Regulations (RIDDOR)	Learning outcomes 1.1, 1.3, 2	Learning outcomes 1.1, 1.3, 2	Topic 2.4
Working at Height Regulations	Learning outcomes 1.1, 6	Learning outcomes 1.1, 6	Topic 2.4
Electricity at Work Regulations	Learning outcome 7	Learning outcome 7	
Personal Protective Equipment (PPE) at Work Regulations	Learning outcomes 1.1, 8	Learning outcomes 1.1, 8	Topic 2.4
Manual Handling Operations Regulations	Learning outcome 5	Learning outcome 5	Topic 2.4
Construction (Design and Management) Regulations	Learning outcomes 1.1, 1.2, 4.1	Learning outcomes 1.1, 1.2, 4.1	Topic 2.4
Construction site health and safety inductions	Learning outcomes 1.2, 1.5	Learning outcomes 1.2, 1.5	
Accidents and emergencies	Learning outcomes 2.2, 3	Learning outcomes 2.2, 3	
Fire safety	Learning outcomes 2.2, 9	Learning outcomes 2.2, 9	
Safety signs and signals	Learning outcomes 1.2, 3.5	Learning outcomes 1.2, 3.5	
Welfare at work	Learning outcomes 1.2, 4	Learning outcomes 1.2, 4	
First aid	Learning outcome 2	Learning outcome 2	

1 HEALTH AND SAFETY ROLES AND RESPONSIBILITIES

Health and safety **legislation** exists to protect us all. It tells us what we should do to keep safe and gives details of different people's roles and responsibilities. Everyone must stick to the guidance. It is essential that you fulfil your role and responsibilities for your own health and safety and that of other people.

For construction, health and safety legislation tells us what should and should not be done by employers and employees to stay safe. Breaking these laws can result in criminal proceedings and then fines, businesses being forced to close and prison sentences for employers or employees.

The Health and Safety Executive

The Health and Safety Executive (HSE) is a government-approved organisation that supervises the workplace to prevent death, injury or ill health. The HSE has the power to make employers change working practices so that the workplace is safer for everyone.

The HSE advises, cautions and investigates when required and carries out research into occupational risks. It aims for its actions to be cost-effective, measured and clear so that businesses will follow the rules.

The HSE focuses on reducing work-related illnesses and deaths. It works with everyone to improve existing laws and regulations and suggests new guidance to make the workplace as safe as possible. If it hears about unsafe working practices, it will step in to enforce the Health and Safety at Work Act 1974 (see below) and often makes unannounced workplace visits. It will issue enforcement notices, such as an **improvement notice** or **prohibition notice**, to any business that has unsafe working practices.

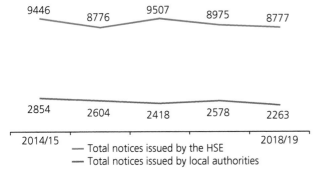

9446 8776 9507 8975 8777

2854 2604 2418 2578 2263

2014/15   2018/19
— Total notices issued by the HSE
— Total notices issued by local authorities

▲ Figure 2.1 Enforcement notices issued by local authorities and the HSE between 2014 and 2019

INDUSTRY TIP

According to the HSE, there were 30 fatalities within the construction sector in 2018/19. This was a significant drop from 49 in 2011/12.

The HSE can be contacted for advice on 0300 790 6787. The number is free to call and lines are open Monday to Friday 8:30 am to 5 pm.

IMPROVE YOUR ENGLISH

An acronym is an abbreviation formed from the initial letters of other words; e.g. HSE stands for Health and Safety Executive.

Research and write out the meanings of the following acronyms:
- PVA
- SBR
- LEV
- MSDs.

2 HEALTH AND SAFETY LEGISLATION

Health and Safety at Work Act 1974 (HASAWA)

The HSE oversees and enforces the Health and Safety at Work Act 1974 (HASAWA). This legislation applies to every kind of workplace and its purpose is to protect all **stakeholders** who might be affected by any type of work being undertaken.

Under HASAWA, the key responsibilities for employers are to provide:
1 a safe system of work
2 a safe place of work
3 safe equipment and machinery
4 safe and competent people working together
5 **risk assessments** and **method statements** (RAMS)
6 information about potential hazards and adequate instruction and training
7 a competent person responsible for health and safety
8 adequate welfare facilities for all stakeholders.

KEY TERMS

Legislation: a law or set of laws suggested by a government and made official by a parliament.

Improvement notice: issued by the HSE or local authority inspector to formally inform a company that safety improvements are needed.

Prohibition notice: issued by the HSE or local authority inspector when there is an immediate risk of personal injury. This is very serious and a company that receives a prohibition notice will clearly be breaking health and safety regulations.

Stakeholder: a person with an interest or concern in a project, especially business.

Risk assessment: the process of identifying hazards and risks that could cause harm.

Method statement: a document to help manage work and ensure that everyone has been told about taking precautions. It often includes a logical sequence of work.

HASAWA lists employees' key responsibilities as:

1 taking care of themselves and others
2 co-operating with their employer in all areas of health and safety
3 not obstructing or interfering with health and safety measures
4 attending all mandatory health and safety training.

Control of Substances Hazardous to Health (COSHH) Regulations 2002

COSHH is the legislation that controls how employers use substances which are hazardous to health in the workplace. In construction, these substances are:

- chemicals
- products that have chemicals added to them
- fumes
- dusts
- vapours
- gases
- biological agents.

What make a substance hazardous?

Under the COSHH regulations, employers are required by law to prevent and reduce any type of exposure to hazardous substances to prevent ill health to their workforce. 'Hazardous substances' include simple substances that could cause breathing problems or spread bacteria, for example, industrial paint which might contain **volatile organic compounds (VOCs)** or mould and algae growth on substrates.

Substances are considered in the form in which they might be used in work activities.

- A hazardous substance might be a chemical compound (a substance with different chemically bonded elements) or it might be a mixture of compounds, micro-organisms or natural materials, such as flour, stone or wood dust.
- Dust of any kind can be hazardous to health under COSHH when it is present at high concentrations in the air.

Gases and vapours

COSHH covers gases and vapours which make it difficult for people to breathe. If a gas or vapour is present at high concentrations in the air, it will replace the oxygen we need to breathe. Many gases are invisible and do not carry a smell, so they are difficult to detect. The oxygen content of the air must be monitored to make sure that gases and vapours do not pose a risk to the health of employees.

Risk of fire

Flammable gases and vapours also carry a risk of fire or explosion. Employers might need to carry out additional duties to keep employees safe from such substances. These duties are set out in the Dangerous Substances and Explosive Atmospheres Regulations 2002 (DSEAR) and the Confined Spaces Regulations 1997.

Biological agents

Biological agents can also be considered to be hazardous substances:

1 Micro-organisms such as bacteria, viruses and fungi and the agents that cause diseases, such as:
   - Weil's disease (leptospirosis) – often present in cow and rat urine and can contaminate fresh water supplies
   - bovine spongiform encephalopathy (BSE, also known as mad cow disease) – can be present in the food chain, mainly from beef stock.
2 Damp spores caused by poor ventilation are common in construction. If they are not dealt with correctly, they can transmit to humans and cause breathing and respiratory problems.
3 Material which contains infectious agents, such as contaminated water or waste material.
4 Substances that cause occupational asthma if a person might develop the symptoms of asthma, such as wheezing and shortness of breath, after coming into contact with the substance within the workplace.

KEY TERMS

Volatile organic compounds (VOCs): organic chemicals that have a high vapour pressure at room temperature, including human-made and naturally occurring chemical compounds. Nearly all scents and odours are classed as VOCs.

Biological agents: types of bacteria, virus, protozoan, parasite or fungus.

ACTIVITY

In small groups:
- research what is meant by volatile organic compounds (VOC) and local exhaust ventilation (LEV)
- give some examples of their uses within construction.

COSHH risk assessment

This is a document to examine tasks and processes that use any potential hazardous substance. Its purpose is to identify hazardous substances an operative might have to use and the precautions that must be taken to avoid harm or ill health from contact with these substances.

There are three ways in which hazardous substances can enter the body:

1 inhalation – breathing in a substance
2 absorption – a substance entering the body through the skin
3 ingestion – taking in the substance through the mouth.

INDUSTRY TIP

COSHH data sheets provide information on products to help users make a risk assessment and describe any hazards the substance presents. They contain information on handling, storage and any emergency measures to be taken in case of accident or misuse.

ACTIVITY

Research the precautions and safety measures that will need to be in place when working with polyvinyl acetate (PVA) and styrene butadiene (SBR). Find reliable information by searching online for COSHH data sheets.

Provision and Use of Work Equipment Regulations 1998 (PUWER)

The PUWER regulations aim to make the workplace safer for anyone who uses or comes into contact with machinery and equipment. The regulations ensure that machinery and equipment are:
- suitable for purpose
- maintained regularly and are in good working order
- operated by people who have been trained to use them correctly
- inspected by a competent worker who keeps a maintenance log.

PUWER regulations apply to all businesses where equipment is operated. However, they do not apply to those who have sold or supplied any equipment – the responsibility for the equipment rests with the employer. This means that, when buying equipment, it is important to check whether it will be safe to use or will need to be adapted for use in the workplace.

ACTIVITY

Research how electrical equipment is maintained and proven to be safe to work with in the workplace under PUWER.

The Control of Noise at Work Regulations 2005

While working in the construction workplace, it is highly likely that you will be exposed to noise. These regulations protect against excessive exposure to noise which could result in damage to hearing, such as hearing loss or tinnitus (permanent ringing in the ears).

1   Damage might be caused by the volume of noise (how loud it is), which is measured in decibels. Examples of decibel levels include:
   ● a normal conversation: 60 decibels
   ● a working lawn mower: 90 decibels
   ● a loud rock concert: 120 decibels.
2   Being exposed to excessive noise for long periods of time on a regular basis can also cause damage. The regulations state the maximum periods of time a person can safely be exposed to particular decibel levels.

▲ Figure 2.2 Ear defenders

▲ Figure 2.3 Ear plugs

The Regulations state the following:
● If the daily or weekly average exposure to noise is 80 decibels, the employer must provide information and training.
● If the daily or weekly average exposure to noise is 85 decibels, employers must provide hearing protection.
● The upper exposure limit is 87 decibels. This means workers must not be exposed to this level of noise.

Reporting of Injuries, Diseases and Dangerous Occurrences Regulations 2013 (RIDDOR)

These regulations state that employers must report serious workplace accidents, illnesses caused by the workplace and dangerous incidents (near misses). The regulations apply to all work premises. Incidents should be reported to the HSE and not reporting an accident is breaking the law.

Under the regulations, all of the following must be reported:
● deaths and injuries caused by workplace accidents
● occupational diseases
● **carcinogens** (such as asbestos), **mutagens** (such as gas particles in the soil discovered during excavations) and biological agents
● specified injuries to workers
● dangerous occurrences
● gas incidents.

When companies report these accidents, the HSE can look for trends of where and how any risk may occur. This helps them to know whether they should investigate and to identify ways of preventing future accidents.

KEY TERMS

Carcinogens: substances which can cause cancer.
Mutagens: agents such as radiation or chemical substances which can cause genetic mutation in the body.

ACTIVITY

1   Find out why electric leads should not trail on the floor on construction sites.
2   Describe in detail the action you would need to take if you came across electric leads on the floor.

The Working at Height Regulations 2005

The purpose of the Working at Height Regulations 2005 is to prevent death and injury caused by a fall from height. Working at height means work in any place where a person could fall a distance likely to cause personal injury, if there were no precautions in place. For example, you are working at height if you:

- are working on a ladder or flat roof
- could fall through a fragile surface
- could fall into an opening in a floor or a hole in the ground.

The term 'working at height' refers to any height off the ground, from a few centimetres to many metres high. If an operative falls from ground level into a trench, this is still considered a fall from height.

All employers or anyone else in control of a working at height activity must make sure that this kind of work is properly planned, supervised and carried out by competent people, using the right type of equipment. A risk assessment should be carried out before any work at height is undertaken. Employees must take reasonable care of themselves and others who are working at height and employers and employees should co-operate to meet all health and safety duties and requirements.

ACTIVITY

In small groups:

- research what is meant by 'soft landing systems'
- find or produce some images of these systems, and discuss their use.

It is a common occurrence for a construction worker to work high off the ground, for example when:

- using a hop-up
- using a ladder
- using and working on scaffolding
- working on roofs
- working on chimneys.

The Working at Height Regulations state that operatives must do everything possible to reduce risk of injury or death from working at height. Workers must undertake training, report any hazards to a supervisor/line manager and use all necessary and available safety equipment.

Before working at height, these precautions should be taken:

- If the work does not need to take place at height, avoid doing so.
- If the work has to be done at height, take precautions to prevent falls, either by using an existing place of work that is already safe or by using the right type of equipment.
- Assess what would happen if there was a fall from height and use the correct type of equipment to try to prevent this from happening.

Ways of reducing the risk from working at height include:

- doing as much work as possible from the ground
- making sure that workers can get safely to and from their position working at height
- maintaining the equipment to be used and checking that it is suitable for the job
- not overloading or overreaching
- taking precautions when working on or near fragile surfaces
- providing protection from falling objects
- considering emergency evacuation and rescue procedures.

INDUSTRY TIP

One of the most common causes of accidents when working at height is a fall from a roof, through a fragile roof or a roof light. These falls can result in death or serious injury and can occur on construction sites or when roof repair work or cleaning is being carried out. These accidents are preventable and information on safe working practices is available from the HSE.

Using ladders

Using a ladder might appear to be dangerous but this is allowed by HASAWA and ladders can be a practical choice for some tasks. The right type of ladder should be used and operatives must know how to use the ladder safely.

A risk assessment should take place before a ladder is used for work at height. If the assessment identifies that the task involves staying up a leaning ladder or stepladder for more than 30 minutes at a time, other safer equipment should be used. A ladder should only be used where it will be level and stable and can be secured.

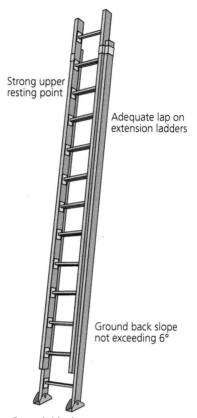

Strong upper resting point

Adequate lap on extension ladders

Ground back slope not exceeding 6°

Ground side slope not exceeding 16°, clean and free of slippery algae and moss

▲ Figure 2.4 Using a ladder safely

▲ Figure 2.5 Resting ladders on plastic guttering can cause it to bend and break

The Electricity at Work Regulations 1989

The Electricity at Work Regulations ensure that work involving electricity or electrical equipment is carried out safely. Under their duty of care, employers must make sure that any electrical systems and equipment are safe and regularly maintained.

If at all possible in construction, it is safer to use battery-powered equipment. Mains power is 230 volts (v) – this can be dangerous and an electric shock might result in bad burns or death.

For safer use on site, a transformer using 110 V can be used. On larger sites it might be necessary to use a large generator capable of 400 V, so that it is safe to plug in more than one item at once. All three voltages of electricity on site are colour-coded: yellow for 110 V, blue for 230 V and red for 400 V.

INDUSTRY TIPS

Never rest an aluminium extension ladder on a plastic-type gutter as your body weight might crush the gutter, causing damage to it and probably a fall from height. Avoid this by using a ladder stand off to work safely. This piece of equipment will sit on the masonry substrate to stop the ladder leaning on the gutter. This makes it safer to work off the ladder, reducing the chances of it sliding horizontally and causing a fall.

Health and safety legislation is put in place to protect you and others. If you do not follow the law, you are placing everyone at serious risk.

- When using a ladder, it should be at 75°.
- Use the 1 in 4 rule: 1 unit out for every 4 units up.

▲ Figure 2.6a 110 V 1 phase – yellow

▲ Figure 2.6b 230 V 1 phase – blue

▲ Figure 2.6c 400 V 3 phase – red

Portable appliance testing

Portable appliance testing (PAT) is the examination of electrical appliances and equipment to ensure they are safe to use. All portable equipment should be maintained to make sure that it is safe. Some of this testing will be visual, for example, looking at equipment to check that it is not damaged. However, some damage will not be visible and formal testing by a trained person will be necessary.

The amount of testing required will depend on the type of equipment. Tools used on a construction site will need to be tested regularly to make sure they remain safe to use.

It is useful to label equipment to show when it was last tested. This will help to make sure that equipment is regularly tested. However, there is no legal requirement for this.

New equipment does not need to be tested, but should be given a visual check before use to make sure it is not damaged.

Personal Protective Equipment (PPE) at Work Regulations 2002

PPE is equipment that will protect the user against health or safety risks at work. It can include items such as safety helmets, gloves, eye protection, high-visibility clothing, safety footwear and safety harnesses. It also includes respiratory protective equipment (RPE).

PPE helps to protect operatives against accidents or injury and should be used alongside other measures to keep people healthy and safe in the workplace. It is necessary even where other safe systems of work are in place, to protect from hazards such as:

- breathing in contaminated air, which can damage the lungs
- falling materials
- flying particles or splashes of corrosive liquids which can irritate the eyes
- contact with corrosive materials which can damage the skin
- extreme temperatures, either cold or hot.

▲ Figure 2.7 Workers wearing safety harnesses on an aerial access platform

The PPE regulations place a duty on:

- the makers of PPE to ensure that it satisfies basic health and safety requirements
- employers to provide PPE and make sure that operatives are trained to use it correctly.

PPE should be checked before use and discarded if damaged.

Other special regulations cover contact with hazardous substances (including lead and asbestos) and also noise and radiation.

Selection and use

When selecting what items of PPE are required, you will need to know which risks the user will be exposed to and how long they will be exposed for. You will also need to bear in mind these details:

- Products which meet the standards of these regulations will be marked 'CE'.
- The size, fit and weight of the PPE: well-fitting PPE will keep the user safe while carrying out their work.
- If more than one item of PPE is required, make sure they can be used together; for example, wearing safety glasses might disturb the seal of a respirator, causing an air leak.

- Users must be trained in how to use the PPE safely; for example, how to remove gloves without contaminating the skin.
- The use of PPE should be encouraged; for example, through the use of safety signs in the workplace.

PPE should be properly stored and maintained to make sure it is safe to use. Replacements should be available in case a piece of PPE is damaged and there should be enough PPE equipment for all people who require it, including anyone who might not be a regular user (such as a visitor).

There should be a designated person who is responsible for the storage and maintenance of PPE, but you are responsible for reporting any faults and using the PPE properly.

Types of PPE

Type of PPE	Hazards	Options for use and considerations
Knee pads: used when performing kneeling activities, to protect from compression and contamination injury	<ul><li>Kneeling on debris</li><li>Activity taking a long time</li><li>Chemical reactions (e.g. lime burns)</li></ul>	<ul><li>Knee pads within protective clothing trousers</li><li>Knee pad inserts for work trousers</li><li>Knee pads with Velcro or belt and buckle fit</li></ul>
Hi-viz clothing: used as standard so that operatives are clearly visible when engaging in work activities	<ul><li>Moving plant/construction traffic</li><li>Busy pedestrian routes</li></ul>	Different forms of hi-viz clothing (e.g. vest, trousers, jackets)
Skin protection, such as sun cream or barrier cream: to protect from harmful sun rays and harmful substances	Sun burn and skin damage (e.g. dermatitis)	<ul><li>High-factor sun block</li><li>Three-stage skin protection station (see page 46)</li></ul>

Type of PPE	Hazards	Options for use and considerations
Waterproof clothing: to protect from poor weather conditions (e.g. wind, rain and snow)	Getting wet, cold and hypothermic	Breathable waterproof clothing (e.g. trousers, jackets, hats, gloves, socks)
Safety glasses: always use on site and in college workshops to protect from plaster/lime sand mortar splashes	● Chemical or molten metal splash ● Dust ● Projectiles ● Gas and vapour ● Radiation	● Safety spectacles, goggles, face screens, face shields, visors ● Make sure the eye protection chosen provides protection against all likely hazards of the task and fits the user properly
Hard hat: always used on site to protect from falling objects or minor head collisions	● Impact from falling or flying objects ● Risk of head bumping, hair becoming tangled in machinery, chemical drips or splashes, change of weather or temperature	● Industrial safety helmets, bump caps, hairnets and firefighters' helmets ● Some safety helmets incorporate or can be fitted with eye or ear protection ● Neck protection should also be considered (e.g. scarves for use during welding)
Ear defenders: always used on site to protect from excessive noise caused by machinery, saws and drilling	● A combination of sound level and duration of exposure ● Very high-level sounds are a hazard even with short duration	● Earplugs, earmuffs, semi-insert/canal caps ● The right hearing protection should be used for the type of work and noise that the user will be exposed to ● Protectors should reduce noise but also allow for communication between users
Gloves: always used on site and sometimes in college workshops to protect from abrasions	● Abrasion, temperature extremes, cuts and punctures, impact, chemicals, electric shock, radiation, vibration, biological agents and prolonged immersion in water ● Gloves might not be suitable for operating some machinery (e.g. bench drills) as the gloves might get caught	Gloves, gloves with a cuff, gauntlets and sleeving that covers part of or the entire arm

Type of PPE	Hazards	Options for use and considerations
Safety boots: always worn on site and in college workshops to protect from impact and falling debris	• Wet, hot and cold conditions • Electrostatic build-up • Slipping, cuts and punctures • Falling objects • Heavy loads • Metal and chemical splashes • Vehicles	• Footwear with protective toecaps and penetration-resistance; wellington boots; specific footwear for particular work areas and equipment (e.g. foundry boots and chainsaw boots); oil- or chemical-resistant soles • Appropriate footwear should be selected for the risks identified
Dust mask: always used on site and sometimes in college workshops to protect from the inhalation of dust and toxic fumes	• Oxygen-deficient atmospheres • Dusts, gases and vapours	• Respiratory protective equipment (RPE) – respirators such as simple filtering face pieces and power-assisted respirators, which must fit properly to be effective • Breathing apparatus – fresh-air hose, compressed airline and self-contained breathing apparatus
Whole body protection	• Heat, chemical or metal splashes • Spray from pressure leaks or spray guns • Contaminated dust • Impact or penetration • Excessive wear or entanglement of own clothing	• Conventional or disposable overalls, boiler suits, aprons, chemical suits • Materials used might be flame-retardant, anti-static, chain mail, chemically impermeable and hi-viz • Other protection such as safety harnesses or life jackets might be needed

Emergency equipment

Careful selection, maintenance and regular training is needed for equipment used in emergencies, such as compressed-air escape breathing apparatus, respirators and safety ropes or harnesses.

INDUSTRY TIP

Hard hats have an expiry date and might not offer the level of protection required if used after this date. The expiry date is usually stamped below the brim.

If your hard hat has expired, speak to your line manager to request a new one. By law, an employer must provide this free of charge.

ACTIVITY

In small groups, research the cost of these PPE items:

- hard hat
- safety glasses
- protective footwear
- hi-viz vest
- protective gloves
- dust mask.

IMPROVE YOUR MATHS

When you have found the costs of the PPE items in the Activity, work out the **net costing** and the **gross costing** including VAT at the current rate. See page 44 for information on working out VAT. Remember to show your working out.

KEY TERMS

Net cost: value of something after taxes and other costs have been deducted.

Gross cost: value of something including taxes and other costs.

The Manual Handling Operations Regulations 1992

Manual handling covers a wide variety of activities, including lifting, lowering, pushing, pulling and carrying. If you do not carry out these tasks correctly, you could be injured. Possible injuries include musculoskeletal disorders (MSDs), such as pain in and injuries to arms, legs and joints and repetitive strain injuries. These injuries can be caused by any work which involves heavy manual labour, awkward postures and repetitive movements.

The Manual Handling Operations Regulations 1992 set out measures to deal with the risks from all work which involves manual handling.

Manual handling injuries can have serious implications for the employer and the person who has been injured. To help prevent manual handling injuries, employers must look at the risks of tasks that involve work with heavy loads and establish sensible health and safety measures to prevent injury.

When assessing the possibility of manual handling injuries in the workplace, you must consider:
- what the individual user is able to lift
- the types of loads that need to be moved
- the working conditions
- what training is required.

If a load needs to be moved manually, these measures can be taken to help avoid injuries:
- Reduce the amount of twisting, stooping and reaching required.
- Do not lift a load from floor level or above shoulder height, especially heavy loads.
- Arrange storage areas to minimise the need to carry out such movements. Heavy items could be delivered directly or closer to the storage area to minimise the amount of manual handling required.
- Reduce the distances which the load needs to be carried as much as possible.
- If possible, break down the weight into smaller units so the individual loads are less heavy.

Lifting techniques

If a load needs to be lifted and moved, the following should be considered before the process begins:
- Identify any handling aids that could be used.
- Remove any obstructions between the place where the load is stored and the place it needs to be moved to.
- Make sure the operator knows where the load should be placed, so it does not need to be moved again.
- Where the load will need to be lifted for a long time, arrange for somewhere to rest midway between the beginning and end, such as a table or bench. This will allow the operative handling the load to rest and if necessary change their grip.

During the lift, the HSE recommends a number of techniques to help avoid injury, as shown in the following table.

▲ Figure 2.8 An operative lifting heavy bricks safely

Technique	Explanation
Adopt a stable position	Your feet should be apart with one leg slightly forward (or beside the load, if it is on the ground) to maintain balance. Be prepared to move your feet when lifting to maintain stability. Avoid tight clothing or unsuitable footwear, which may make this difficult.
Get a good hold	Where possible, hug the load close to your body. This may be better than gripping it tightly with only your hands.
Start in a good posture	Before you start lifting, slightly bend your back, hips and knees. This is preferable to fully flexing your back (stooping) or fully flexing your hips and knees (squatting).
Do not flex your back any further while lifting	This can happen if you begin to straighten your legs before you start to raise the load.
Keep the load close to your waist	Keep the load close to your body for as long as possible while lifting. Keep the heaviest side of the load next to your body. If a close approach to the load is not possible, try to slide it towards your body before attempting to lift it.
Avoid twisting your back or leaning sideways, especially while your back is bent	Shoulders should be kept level and facing in the same direction as your hips. Turning by moving the feet is better than twisting and lifting at the same time.
Keep your head up when handling	Look ahead, not down at the load, once you are holding it securely.
Move smoothly	The load should not be jerked or snatched as this can make it harder to keep control and can increase the risk of injury.
Do not lift or handle more than you can easily manage	There is a difference between what people can lift and what they can safely lift. If in doubt, seek advice or get help.
Put down, then adjust	If precise positioning of the load is necessary, put it down first, then slide it into the desired position.

INDUSTRY TIP

When using a lifting aid, you are still manual handling. However, lifting aids reduce the risk of many potential injuries such as back strains, hand and feet crush or abrasion injuries and muscle strains.

▲ Figure 2.9 Mechanical lifting aid pump truck

Construction (Design and Management) Regulations 2015

These regulations (known as CDM 2015) govern the way in which all sizes and types of construction projects in the UK are planned. The main set of regulations is for managing the health, safety and welfare of construction projects. It applies to all construction work including new build, demolition, refurbishment, extensions, conversions and any repair and maintenance works.

Under the regulations, all stakeholders in a project have duties that need to be carried out. The HSE lists these duties, which are summarised in the table below. Remember the following points:

- A duty holder can carry out more than one of these roles if they have the skills, knowledge and experience to do so.
- An organisation can carry out more than one role, if it is able to do so while ensuring the health and safety of everyone involved in the project.
- CDM 2015 only applies to domestic clients if the work is carried out by someone other than the client. If you carry out your own domestic work, CDM does not apply and the work is considered to be 'do it yourself' (DIY).

This table describes the duties of different duty holders under the CDM regulations.

Duty holder	Description	Duties
Commercial clients	Organisations or individuals for whom a construction project is carried out. This is done as part of a business.	Make arrangements for managing a project: ● Appoint other duty holders as appropriate. ● Allocate enough time and resources to the project. ● Prepare relevant information and provide it to other duty holders. ● Ensure that the principal designer and principal contractor carry out their duties (see below). ● Provide welfare facilities.
Domestic clients	People who have construction work carried out on their own home (or the home of a family member). This is **not** done as part of a business.	Their client duties are normally transferred to: ● the contractor for single contractor projects ● the principal contractor, if there is more than one ● a principal designer, if the client chooses this (by written agreement).
Principal designers	Designers appointed by the client in projects involving more than one contractor. They can be an organisation or an individual with the knowledge, experience and ability to carry out the role.	Plan, manage, monitor and co-ordinate health and safety before construction begins. This includes: ● identifying, eliminating or controlling foreseeable risks ● ensuring that designers carry out their duties ● preparing and providing relevant information to other duty holders ● liaising with the principal contractor in the planning, management, monitoring and co-ordination of the construction phase.
Principal contractors	Contractors appointed by the client to co-ordinate the construction phase of a project where it involves more than one contractor.	Plan, manage, monitor and co-ordinate health and safety in the construction phase of a project. This includes: ● liaising with the client and principal designer ● preparing the construction phase plan ● organising co-operation between contractors and co-ordinating their work. During construction, the principal contractors must ensure that: ● suitable site inductions are provided ● reasonable steps are taken to prevent unauthorised access ● workers are consulted and engaged in securing their health and safety ● welfare facilities are provided.
Workers	Those working for or under the control of contractors on a construction site.	Workers must: ● be consulted about matters which affect their health, safety and welfare ● take care of their own health and safety and of others who might be affected by their actions ● report anything they see which is likely to endanger either their own or others' health and safety ● co-operate with their employer, fellow workers, contractors and other duty holders.

▲ Figure 2.10 A client, a contractor and an operative looking over building plans ahead of construction

Health and Safety Executive

Construction Phase Plan (CDM 2015)
What you need to know as a busy builder

Under the Construction (Design and Management) Regulations 2015 (CDM 2015) a **construction phase plan** is required for every construction project. This does not need to be complicated.

If you are working for a domestic client, you will be in control of the project if you are the only contractor or the principal contractor.

You will be responsible for:
- preparing a plan;
- organising the work; and
- working together with others to ensure health and safety.

You could be a builder, plumber or other tradesman, doing small-scale routine work such as:
- installing a kitchen or bathroom;
- structural alterations, eg chimney breast removal;
- roofing work, including dormer windows;
- extension or loft conversion.

A **simple plan** before the work starts is usually enough to show that you have thought about health and safety.

If the job will last longer than 500 person days or 30 working days (with more than 20 people working at the same time) it will need to be notified to HSE and it is likely to be too complex for this simple plan format.

The list of essential points below will help you to **plan** and **organise** the job, and **work together** with others involved to make sure that the work is carried out without risks to health and safety. It will also help you to comply with CDM 2015. You can use the blank template on page 2 to record your plan.

Plan	Working together
Make a note of the key dates, eg: ■ when you'll start and finish; ■ when services will be connected/disconnected; ■ build stages, such as groundwork or fitout. You will need to find out information from the client about the property, eg: ■ where the services and isolation points are; ■ access restriction to the property; ■ if there is any asbestos present.	It may be useful to record the details of anybody else working on the job, including specialist companies and labourers. Explain how you will communicate with others (eg via a daily update), provide information about the job, coordinate your work with theirs and keep them updated of any changes, eg: ■ to site rules; ■ to health and safety information; ■ what you will do if the plan or materials change or if there are any delays; ■ who will be making the key decisions about how the work is to be done.

Organise

- Identify the main dangers on site and how you will control them, eg:
  - the need for scaffolding if working at height;
  - how structures and excavations will be supported to prevent collapse;
  - how you will prevent exposure to asbestos and building dust;
- how you will keep the site safe and secure for your client, their family and members of the public.
- Make sure that there are toilet, washing and rest facilities.
- Name the person responsible for ensuring the job runs safely.
- Explain how supervision will be provided.

If you are unsure about how you can make your site safer, see www.hse.gov.uk/construction for more information and to download other Busy Builder sheets. See www.citb.co.uk for a free smartphone app *CDM wizard*.

1

▲ Figure 2.11 Example of a construction phase plan

Control of Vibration at Work Regulations 2005

These regulations cover hand–arm vibration at work. Hand–arm vibration is vibration transmitted into the hands and arms when using hand-held power tools and equipment and is a risk if these tools are used for a long time. Examples of tools used in plastering operations which have potential to cause hand–arm vibration are:

- paddle mixers (whisks)
- projection plastering machines
- SDS drills
- mechanical breakers
- power floats
- auto-feed screwdrivers
- impact drivers
- multi-tools.

▲ Figure 2.12 An operative taking a rest from using a power tool

There are two main types of ill-health that can be caused by hand–arm vibration. These are hand–arm vibration syndrome (HAVS) and carpal tunnel syndrome (CTS).

The main effects of HAVS are:

- reduced strength in the hands, which can stop you from working safely and carrying out tasks involving fine work
- painful fingers, reducing your ability to work in cold or damp conditions, such as outdoors.

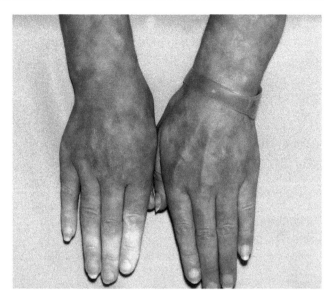

▲ Figure 2.13 The effect of HAVS

The effects of CTS include:

- tingling, numbness, pain and weakness in the hand: this can interfere with work and everyday tasks and might affect your ability to work safely.

Symptoms of both syndromes might come and go, but if you are exposed to vibration for a long time, the symptoms could worsen and become permanent, causing pain, distress and sleep disturbance.

To avoid permanent damage, follow these guidelines:

- Always use correct PPE and take regular breaks when using these types of tool.
- Do not use these tools for long periods of time.
- Attend training on the risks and correct use of this equipment and remember the safety information.

③ CONSTRUCTION SITE HEALTH AND SAFETY INDUCTIONS

A construction site induction is a safety briefing given to workers at the start of any new work. The induction will explain the safety rules and controls in place, any potential hazards that workers might be exposed to and how to work safely on the site. Operatives will not be allowed to start any works until the induction process is completed.

Operatives will also need a current **CSCS (Construction Skills Certificate Scheme) card** to begin working on many sites. To gain a CSCS card, you need to provide proof of your qualifications and take an online health and safety test.

CSCS cards are valid for five years. After this time, you have to retake the health and safety test and update your qualifications to gain a renewed card.

KEY TERM

CSCS (Construction Skills Certification Scheme) card: provides proof that individuals working on construction sites have the appropriate training and qualifications for their on-site job role.

INDUSTRY TIP

For trade areas, operatives now need to have the highest trade area NVQ qualification attached to the CSCS card. Apprentices will be exempt, but must be working towards these qualifications.

ACTIVITY

Research the CSCS scheme and find out what you need to do to gain a CSCS card.

Construction site induction

The purpose of a construction site induction is to inform workers about the organisation and operation of the site and about their individual responsibilities. Construction site inductions are required by law to ensure a safe site.

It is standard procedure to receive an induction on your first day on site. You will learn about safety rules and controls, any hazards you might be exposed to, how to work safely on site and the provisions for first aid. No two sites are the same, so every worker should have a suitable site induction on every site they work on.

Giving vital safety information during an induction could be the difference between operating a safe site and a dangerous one. If all workers know what to do

and what to expect, they will be able to work safely together.

The principal contractor is responsible for all site inductions. On smaller projects there might not be a principal contractor, but it is still a legal requirement under CDM 2015 to provide an induction to any workers including sole contractors (working on their own). Whenever a new worker starts on a site, there must be an induction, even if it is for only one person.

ACTIVITY

- Research CDM 2015.
- List four important responsibilities listed in these regulations.

INDUSTRY TIPS

The principal contractor has overall control of the construction process when there is more than one contractor working on site. They are appointed by the client to plan, manage and monitor all aspects of the development.

Working on your own is sometimes unavoidable. When this happens, a Lone Worker Risk Assessment must be completed to assess the risk of illness, injury or attack. A supervisor will usually make sure a lone worker has a mobile phone and will check on them every 30 minutes or so.

HEALTH AND SAFETY

Always check with your line manager/supervisor if there is anything you do not understand from a health and safety perspective. If you are not sure about something, do not continue to work.

Construction site toolbox talks

A toolbox talk is a short presentation to a site's workforce on a single aspect of health and safety. The name comes from the team gathering for a discussion around a toolbox on a site, but these meetings can be held anywhere suitable in the workplace.

A toolbox talk:

- allows management and workers to explore the risks of any health and safety issues on site and think about ways to deal with them
- is usually a brief meeting (about 15 minutes), carried out on site before works begin
- is a good way to reinforce safety basics and inform workers about any changes to the site or working conditions since they were last on site
- is not a legal requirement, but is regarded as best practice.

INDUSTRY TIP

Toolbox talks are important because they:
- promote safety in the workplace
- keep workers informed and remind them of their role in keeping the site safe
- improve communication and productivity
- function as an updated record of hazards and action plans.

---

4 ACCIDENTS AND EMERGENCIES

In construction there is always the possibility of some sort of accident or emergency occurring, despite many improvements to safety measures implemented over the years. By law, employers must have practical measures in place to help prevent accidents and must have a system for reporting any work-related accidents and events. Under RIDDOR 2013 (see page 66) any injury which disables a worker from their duties for up to seven days must be reported to the HSE. This report must be made within 15 days of the accident.

Major types of emergencies in the workplace include:
- floods
- fires
- toxic gas releases
- chemical spills
- explosions
- radiological accidents.

As part of their health and safety induction, everyone in the workplace should know how to respond if an emergency occurs. The induction should provide information on:

- who to alert when an accident happens and how
- the supervisor for emergency situations, such as a fire marshal
- the available emergency equipment
- where to find first aid kits
- emergency evacuation plans.

There must be a formal recording and reporting system in place for any accidents or near misses that occur in the workplace. Documentation should include an accident book, a near miss reporting system such as a card drop box (this can be anonymous) and first aid records.

All reports should be recorded electronically and backed up. This is important in case other agencies such as the police, fire service or HSE need specific information when investigating. These records are maintained by the company's health and safety officer.

Emergency and accident situations will be dealt with by trained first aiders and supervisors who must have current first aider qualifications. These people are identified during the induction process and often display colour-coded lanyards or hi-viz vests to make them more identifiable.

Hazards in the workplace

Good housekeeping is important and should include the following points to keep the workplace free from hazards:
- clean and tidy work areas and transition areas
- effective RAMS (risk assessments and method statements)
- a system for identifying hazards
- use of skips and chutes
- safe storage of chemicals and combustibles (COSHH)
- segregation of waste materials
- clear, informative signs and notices
- clear access in transitional areas, in particular to fire escapes
- clear access to fire extinguishers.

The construction workplace is likely to include these types of hazard:
- fires
- slips, trips and falls

- hazardous substances (COSHH)
- electricity
- asbestos
- manual handling
- plant and vehicle movement
- adverse weather.

▲ Figure 2.14 Cables can be a trip hazard on site

Around the construction site there will be many signs providing safety information. These are colour-coded to provide different types of information. See page 83 for more information.

Health and welfare

CDM 2015 identifies requirements for adequate welfare facilities, such as washing and toilet facilities and canteen/eating areas (see pages 84–85).

Some operatives might take prescribed medication, for example, for anxiety, diabetes or epilepsy. This could have an effect on their well-being and safety. It is good practice to tell the employer about this before starting on site, so that a safe system of work can be put in place. However, there must be zero tolerance of any operatives who use recreational drugs or alcohol within the workplace, as this will put them and others in danger.

5 FIRE SAFETY

Most fires in the workplace can be prevented, if employers and employees follow the right approaches to fire safety.

The main requirements for fire safety come from the Dangerous Substances and Explosive Atmospheres

Regulations 2002 (DSEAR). These regulations require employers to assess the risk of fires and explosions arising from work activities involving dangerous substances and to eliminate or reduce these risks.

The Regulatory Reform (Fire Safety) Order 2005 covers general fire safety in England and Wales.

> ### INDUSTRY TIP
>
> If you discover a fire at work, you must raise the alarm if safe to do so and leave the building immediately.

General fire safety hazards

For a fire to start, it needs a source of:

- ignition, such as heaters, lighting, naked flames, electrical equipment, smokers' materials (cigarettes, matches etc.) or anything else that can become very hot or cause sparks
- fuel, such as wood, paper, plastic, rubber or foam
- oxygen (this can include air).

Employers or building owners must carry out a fire safety risk assessment. This is similar to a standard risk assessment and can be carried out as part of it. When the site has been risk-assessed, employers/building owners must put appropriate fire safety measures in place to minimise the risk of injury or death if a fire breaks out.

The fire safety risk assessment should identify what could cause a fire to start (sources of ignition that might be present and sources of fuel for a potential fire) and who would be at risk if a fire breaks out. Then action can be taken to avoid these risks, or to reduce and manage them if they cannot be avoided completely due to the type of work being carried out.

To reduce the risk of fire, the site should follow these points:

- Keep sources of ignition and fuel apart from each other.
- Take action to avoid any accidents that might cause a fire; for example, make sure any sources of heat cannot be knocked over.
- Follow the housekeeping rules on page 79 to make sure there is no build-up of any materials that could cause a fire, such as dust, grease or rubbish.
- Employers should install safety equipment such as smoke detectors and fire alarms. These should be

tested regularly and everyone on site should know how they work.

- Fire-fighting equipment should be available and everyone on site should be trained in how to use it.
- Keep fire exits and escape routes clear at all times. Make sure that directions to them are clearly visible.
- All workers should receive appropriate training on procedures they need to follow, including fire drills.
- The fire risk assessment should be regularly reviewed and updated.

Working with dangerous substances that cause fire and explosion

Many substances found in the workplace can cause fires or explosions. These range from the obvious like petrol and welding gases, to less obvious materials such as wood dust or engine oil. Work which involves any materials that could burn or explode is hazardous and could cause serious injury.

The following are ways to help prevent accidental fires or explosions:

- Be aware of substances and materials that might burn or explode. Use information from suppliers to identify which materials might be flammable.
- Find out which processes in the workplace might set these materials alight and who might be at risk.
- Consider what measures are needed to reduce or remove this risk.
- Reduce the amount of flammable/explosive substances stored in the workplace.
- Make sure that flammable or explosive substances are disposed of safely.

The fire and rescue authorities deal with fire safety matters in general workplaces, but the HSE deals with these matters on construction sites. The HSE is responsible for making sure that companies stick to the fire safety rules on storing dangerous substances, although the local authority might also inspect the premises.

INDUSTRY TIP

Fire blankets are normally found in kitchens or canteens as they are good at putting out cooking fires.

▲ Figure 2.15 Fire blanket

▲ Figure 2.16a CO₂ extinguisher

▲ Figure 2.16b Dry powder extinguisher

▲ Figure 2.16c Water extinguisher

▲ Figure 2.16d Foam extinguisher

▲ Figure 2.17 Emergency procedure sign

INDUSTRY TIP

Induction processes should always include an emergency evacuation plan.

ACTIVITY

- Research the job of a fire warden.
- List five of their important duties.

⑥ SAFETY SIGNS AND SIGNALS

Safety signs can be found in lots of different places around employment areas. They are put in place to:

- warn of any hazards
- prevent any accidents
- inform on where things are
- inform on what to do in various areas.

INDUSTRY TIP

Be absolutely sure to read and adhere to any safety signs in the workplace. They have been put up for a good reason.

Employers must provide safety signs wherever there is a risk in the workplace. Signs must be clear, easy to read and easy to follow. They should explain:

- instructions (e.g. a No Access sign for areas where employees should not enter)
- hazards that might be present (e.g. a sign warning of corrosive material)
- instructions for fire exits and equipment.

This applies to all places and activities where people are employed. Signs must be maintained to make sure they are visible and relevant and should be explained to employees as part of a site induction to make sure that they are followed.

Safety signs are colour-coded, with each colour meaning something different:

- green: information (red for fire-related information)
- yellow: warning
- blue: mandatory (things that must be done)
- red edging: prohibition (things that must not be done).

The use of safety signs is regulated by the Health and Safety (Safety Signs and Signals) Regulations 1996. The regulations require employers to provide suitable safety signs in the workplace following the results of risk assessments, to make sure that risks in the workplace are reduced.

▲ Figure 2.18 Safety signs

7 WELFARE AT WORK

Welfare facilities mean any facilities in a workplace that are needed for the well-being of the people working there. These include toilets, rest and changing facilities and somewhere clean to eat and drink during breaks. According to the HSE, employers must, 'so far as is reasonably practicable', provide adequate and appropriate welfare facilities provided for their employees.

Toilet and washing facilities

Adequate toilet and washing facilities should be provided:

- There should be enough toilets and washbasins so that there are not long queues to go to the toilet.
- Separate facilities for men and women should be provided, although some employers might also provide gender neutral toilet facilities.
- Facilities for employees with disabilities should be available.
- The walls should be tiled or covered in material to ensure that the facilities are waterproof.
- Toilet paper should be provided and a means of disposing sanitary products for female employees.

- Hot and cold running water, soap and cleaning agents and a means of drying hands (such as paper towels or a hot air dryer) should all be provided.
- In workplaces where employees will become covered in dirt, shower facilities should be provided.
- The facilities should be kept clean and in good condition, so there should be a system in place for cleaning and regularly replenishing supplies of toilet paper, soap, etc.

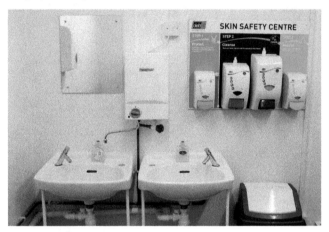

▲ Figure 2.19 Washing facilities

Some worksites might be temporary or in remote locations where new construction is taking place. In these places, flushing toilets and running water need to be provided 'so far as is reasonably practicable'. This might include the use of temporary portable toilets, chemical toilets and washing facilities such as water containers.

This table shows the number of toilets and washbasins that must be provided for different numbers of people on site.

Number of people	Toilets	Washbasins
1–5	1	1
6–25	2	2
26–50	3	3
51–75	4	4
76–100	5	5

Drinking and eating

The law requires employers to provide drinking water that is free from contamination. Water from the public water supply is preferable, but providing bottled water is acceptable.

There should be a suitable area for workers to use during breaks for eating. It should be clean, hygienic and located where food will not become contaminated. Washing facilities should be provided and there should be a means of heating food or water for hot drinks.

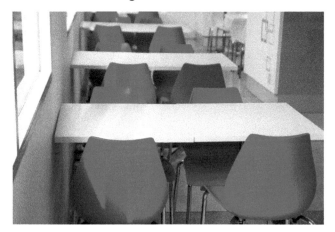

▲ Figure 2.20 Rest areas

Changing facilities

If people will need to change into specialist clothing for work, facilities must be provided to allow this. There should be enough changing rooms for the number of people expected to use them.

Storage facilities should be provided and there should be somewhere to hang clothes, such as hooks or pegs.

Separate changing facilities should be available to men and women, with separate storage for dirty or wet clothes.

Employers should also provide rest facilities for pregnant women and nursing mothers, particularly on larger sites.

INDUSTRY TIP

When attempting to dry clothes on site, do not place them in direct contact with heaters as this is a fire hazard.

▲ Figure 2.21 Lockers must be provided for specialist clothes to be worn at work

Personal hygiene

▲ Figure 2.22 A typical skincare station on a construction site, with applications before work, during work and after work to provide cleanliness and hygiene

When working in construction, you will often become contaminated and dirty from operational tasks. It is important to take personal hygiene seriously:

- You must wash well before, during and after work.
- If possible, use appropriate PPE such as overalls and gloves to protect clothing and skin. This will help to keep vehicles, your home or the office free from building waste and contamination that might be present while you work.
- Every time you attend the welfare facilities and return to the workplace, use the three-point skin care station to protect and clean your hands.

Following these good practices will maintain a safe and hygienic workplace for all stakeholders.

A step-by-step guide to washing hands:

STEP 1 Apply soap to hands from the dispenser.

STEP 2 Rub the soap into a lather and cover your hands with it, including between your fingers.

STEP 3 Rinse hands under a running tap, removing all of the soap from your hands.

STEP 4 Dry your hands using disposable towels. Put the towels in the bin once your hands are dry.

⑧ FIRST AID

Any type of first aid should only be undertaken by someone with adequate training.

HEALTH AND SAFETY

A first aid needs assessment will help employers decide what first aid arrangements are appropriate for their workplace.

As a minimum, a small construction site should have a first aid box and a person appointed to take charge of first aid arrangements, such as:
- looking after all first aid equipment
- keeping check of stocks and rotating items
- calling the emergency services if necessary.

In workplaces where there are more significant health and safety risks, a trained first aider might be needed, although there are no specific rules on how many first aiders are required in the workplace. Where there are special circumstances, such as shift work or sites with several buildings, more trained first aiders might be needed. There should be enough qualified personnel to cover for absences.

Employers must provide information about first aid arrangements to their employees.

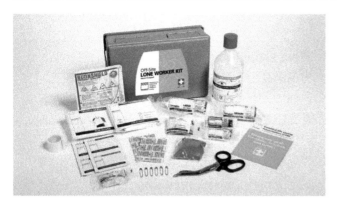

▲ Figure 2.23 A first aid kid

Assessing first aid needs

An assessment of first aid needs will identify the workplace risks and show what first aid provisions need to be put in place. The types of risk will depend on the workplace, workforce and any identifiable hazards. Some considerations are:

- the type of work undertaken and the associated hazards
- the number of employees and types of job role
- the types of injury, accidents or near misses recorded in the past
- the number of buildings and usable working areas (including satellite centres)
- separate needs assessment for satellite centres
- factors such as remote work areas, new start employees, lone workers
- the distribution of workers around the site
- provision for the public and visitors
- the location of buildings and sites to emergency services.

The Health and Safety (First-Aid) Regulations 1981 state that employers must provide appropriate and adequate first aid provisions so that anyone who is injured or taken ill in the workplace receives immediate attention.

ACTIVITY
Your college or place of work will have appointed first aiders and first aid kits. Find out:
- how to identify first aiders
- where the first aid kits are kept
- what should be stocked in the kits.

INDUSTRY TIP

First aid should be carried out by a competent, qualified trained person who will be able to support the patient in an injured state until further treatment is available.

CASE STUDY

Aasma is a plasterer working on a large construction site.

She arrived at her work area to complete some work at height she had started the previous day. She noticed that the mobile tower scaffold she had been using had been moved and that one of the working platforms had been removed and leaned against a nearby wall. The **SCAFFTAG®** she had filled out at the end of the previous day's work had also been removed and discarded on the floor.

Aasma put the mobile tower scaffold back in position, where she had left it the previous day and reinstated the working platform. She also put the SCAFFTAG back on the scaffold. She then checked the scaffold structure for safety, locked the wheels and used the scaffold to complete the job.

KEY TERM

SCAFFTAG®: a scaffold-status tagging system to prevent hazards when working at height and efficiently manage the inspection procedures for scaffolding.

Test your knowledge

1 What does a blue and white safety sign tell you?

   A something you must not do

   B something you must do

   C a general site hazard

   D where the fire exits are

2 Which legislation covers specific guidance on the safe lifting of heavy items on site by operatives?

   A Management of Health and Safety at Work Regulations

   B Manual Handling Operations Regulations

   C Health and Safety at Work Act

   D Construction Design and Management Regulations

3 What must an employer do under the Health and Safety at Work Act?

   A protect the health, safety and welfare of all employees

   B protect the environment during work activities

   C provide transport from employees' homes to the work site

   D ensure employees are paid for the work they do

4 What should you do if you are told that an HSE inspector wishes to interview you after you witnessed an accident?

   A tell the inspector exactly what you saw happen

   B ask your employer what to say

   C say nothing

   D leave the site so you do not have to speak to the inspector

5 Which of the following agencies has a free health and safety advice telephone line?

   A Construction Industry Training Board (CITB)

   B Environmental Protection Agency

   C Royal Society for the Prevention of Accidents

   D Health and Safety Executive (HSE)

6 What should you do if you discover a fire while at work?

   A leave the building straight away

   B raise the alarm if safe, then leave the building

   C carry on working if there is no danger

   D contact the site supervisor and tell them where the fire is

7 What is the reason for recording details of accidents?

   A to keep records for Health and Safety Executive inspectors

   B to keep records for the principal contractor

   C to warn other operatives of the danger

   D to identify the same types of recurring accident

8 Why should you report any near misses?

   A so lessons can be learned to ensure the same thing does not happen again

   B if you do not report a near miss, someone may blame you

   C near misses must always be reported to the principal contractor

   D so that someone can be disciplined for causing the near miss

9 What information should you be told in a site induction?

   A how to perform first aid

   B how to extinguish a fire

   C where you will find and identify first aiders

   D to ensure you have a first aid kit with you at all times

10 What should you read before handling any chemicals on site?

   A the chemical regulations

   B the method statement

   C the COSHH statement

   D health and safety posters on display in the site cabin

11 At what noise level must you wear ear protection?

   A 65 decibels

   B 80 decibels

   C 85 decibels

   D 90 decibels

12 What colour is a CO_2 fire extinguisher?

   A red

   B black

   C yellow

   D blue

13 Which of the following is an example of good housekeeping on a construction site?

   A using skips and chutes

   B going home on time

   C taking regular breaks

   D attending an induction

14 What is the main purpose of a construction site induction?

   A to inform operatives of their responsibilities and the organisation of the site

   B to inform managers of how many personnel are on site

   C to inform workers of who the first aiders are on the construction site

   D to inform all managers and workers of any imminent fire drills

15 What does the acronym HAVS stand for?

   A heavy and vast shipments

   B hand–arm vibration syndrome

   C hospital and visiting statements

   D hand armour vibration scoping

16 When assessing how to lift a load, what must an operative consider?

   A how far the load must be moved

   B how many persons are needed to lift the load

   C what training is required

   D what type of injury an operative may suffer when lifting

17 What essential piece of safety equipment must be used when working on a mobile elevating working platform (MEWP), besides standard PPE?

   A dust mask

   B gauntlet gloves

   C face visor

   D safety harness

18 What does the acronym PAT stand for?

   A problem at towel area

   B portable appliance testing

   C pay and timesheets

   D portable adapted toilet

19 What is the one in four rule for using a ladder safely?

   A 1 in 4 steps must be reinforced

   B check for safety once every four days

   C 1 unit out for every 4 units up

   D clean the ladder once every four weeks

20 What is the main purpose of the Working at Height Regulations 2005?

   A to prevent death and injury caused by falls from height

   B to prove operatives can work safely at height

   C to improve overall health and safety on construction sites

   D to inform managers of any falls from height

CHAPTER 3

INTERNAL PLASTERING AND FIXING DRY LINING

INTRODUCTION

This chapter covers methods of preparing backgrounds, fixing plasterboard, fixing beads and mixing and applying traditional and modern plasters to form and finish one, two and three coat plastering to interior surfaces ready for decoration, to fulfil the customer's needs.

Before you start to apply plaster, you will need to familiarise yourself with the necessary tools. These will be specific to the job. Remember that looking after your tools and keeping them safe and well-maintained will ensure they last for a long time.

You will learn about different techniques and methods required to apply and install internal plastering materials and components to ensure that surfaces are ready for decoration.

Learning about different types of plaster and plasterboard and their properties will help you to understand how the performance of modern buildings has evolved over the years to meet greater demands for thermal, sound, fire, heat and moisture resistance. You will also learn about traditional and sustainable plastering materials and methods, used to preserve our heritage within historic buildings.

By reading this chapter you will gain knowledge of:

1 the tools and equipment required for this work
2 how to prepare the background for plastering
3 how to prepare to fix plasterboard
4 fixing plasterboard by direct bond
5 how to prepare loose plastering materials
6 different plastering systems.

The table below shows how the main headings in this chapter cover the learning outcomes for each qualification specification.

Chapter section	Level 1 Diploma in Plastering (6708-13) Units 121, 122, 123, 124, 125	Level 2 Diploma in Plastering (6708-23) Units 221 and 222	Level 2 Technical Certificate in Plastering (7908-20) Units 202 and 205
Tools and equipment		Unit 221 Learning outcomes 3 and 4	Unit 202 Topics 2.2, 3.1 Unit 205 Topics 1.5, 3.1
Prepare the background for plastering	Unit 121	Unit 221 Learning outcome 5.1	Unit 202 Topics 2.1, 2.3
Prepare to fix plasterboard	Unit 123	Unit 221 Learning outcomes 1 and 2 Unit 222 Learning outcomes 1, 2, 3, 4	Unit 202 Topics 1.1, 1.2, 3.2 Unit 205 Topics 1.1, 1.2
Fix plasterboard by direct bond		Unit 222 Learning outcomes 5, 6	Unit 205 Topics 1.3, 1.4, 2.1, 2.2, 2.3

Chapter section	Level 1 Diploma in Plastering (6708-13) Units 121, 122, 123, 124, 125	Level 2 Diploma in Plastering (6708-23) Units 221 and 222	Level 2 Technical Certificate in Plastering (7908-20) Units 202 and 205
Prepare loose plastering materials for mixing		Unit 221 Learning outcome 5.2	Unit 202 Topic 3.3
Apply plastering systems	Units 122, 124, 125	Unit 221 Learning outcomes 5.3, 5.4, 6	Unit 202 Topic 3.4 Unit 205 Topic 3.2

1 TOOLS AND EQUIPMENT

This table describes the use of some common tools and equipment.

Tool	Use
Hawk	The plasterer's hawk is used to hold and transfer a workable amount of plaster from the spot board to the wall. The hawk is used with the trowel to manipulate and apply the plaster directly on to the background surface. Some plasterers prefer hawks with detachable handles as they are easier to store. Hawks were traditionally made from timber, but modern hawks are made from polyurethane or aluminium.
Laying trowel	The best type of trowel used for applying plaster should be made of stainless steel as this will not rust. Once broken in, the trowel can be used for laying finishing plaster. The trowel should be used to apply undercoat plaster for several weeks or months until the edges become sharp and the corners slightly rounded. Some laying trowels can be purchased pre-worn so they take less time to wear down. In contrast, the blade of the finishing trowel should be kept firm and straight. When purchasing a trowel, ensure that it has a long shank on its rear; this will provide stability for the blade and prevent it from going out of shape.
Gauging trowel	This tool has many purposes and uses within the plastering trade. Its main purpose was to gauge small quantities of plastering materials. Today, this trowel is used for reaching into awkward areas where a normal trowel cannot reach.
Bucket trowel	This is used for cleaning the rim of buckets, but can also be used for cleaning and removing excess material off the straight edge and scraping the floor to clean up plaster droppings. A bucket trowel is also used to transfer mixed material from the bucket onto the spot board.

Tool	Use
Comb scratcher 	A comb scratcher can be used to key the surface when you apply a scratch coat in preparation for the floating coat.
Straight/feather edge 	A straight edge is used for ruling surfaces, checking for straightness or forming the hard angle of a return. It is also known as a feather edge rule because it has a taper on one side, allowing you to **rule** from the tight angles on a wall. Some straight edges can be square on both sides and are generally used for floor screeding or when dry lining.
Darby 	Another tool used for ruling and flattening walls to a smooth surface. You can also use this tool to form returns when applying a scratch coat.
Flexi finishing trowel (flexi trowel, flexi trowel blade) 	Flexi finishing trowels are now used more often because they speed up some aspects of plastering and help to achieve a consistently flat and smooth finish.
Tin snips 	Tin snips are used for cutting various types of trim such as angle beads, stops and rolled **expanded metal lath (EML)** before it is fixed onto the wall plate. Always wear protective gloves when cutting EML because it is very sharp.
Spirit level 	This can be used on its own or with a straight edge, which extends its length. It is used for plumbing and levelling surfaces such as standard angle beads to window openings and returns, or for plumbing dots to form accurate screeds when floating.
Float 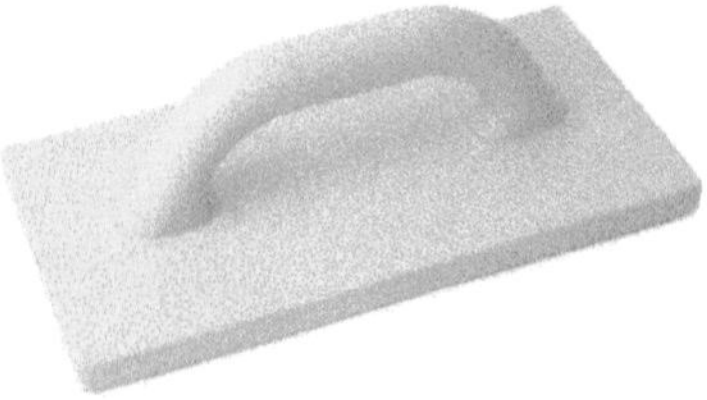	Used for consolidating undercoat surfaces to form either a plain smooth finish or a **key**, preparing for the setting coat. These floats are generally made from polyurethane.

Tool	Use
Devil float 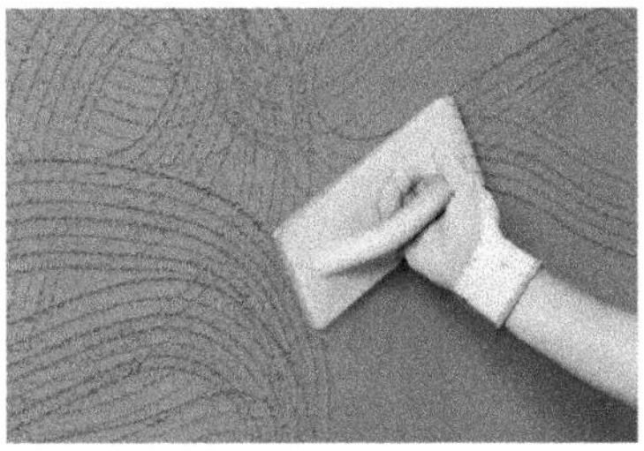	A devil float is made by nailing tin tacks to one end of a float edge. It is used for devilling or keying a surface. When making a devil float, space the tin tacks equally and start from the middle, fixing every 15 mm. Use fixings such as galvanised nails: screws will cause damage and split the edge of the float.
Small tool 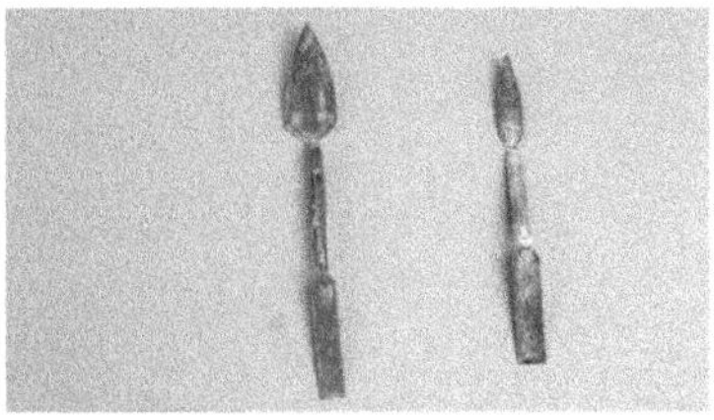	Used in tight, difficult angles. The types shown here are leaf and square.
Mechanical drum mixer 	A mechanical mixer is best used for mixing cement-based plasters. This type of mixing is carried out outdoors, as it can be noisy and the materials used will create a lot of dust.
Plasterer's wheel (plunger) 	A hand-mixing tool used for mixing setting plaster. During and after the mixing process, it should be kept off the floor to prevent any bits of debris (dust and grit) from sticking to the bottom of the tool, which could contaminate the next mix.
Drill and whisk 	A mechanical mixing tool used for mixing lightweight undercoat and setting plasters. It is a fast and efficient way of mixing lightweight plasters.
Flat brush 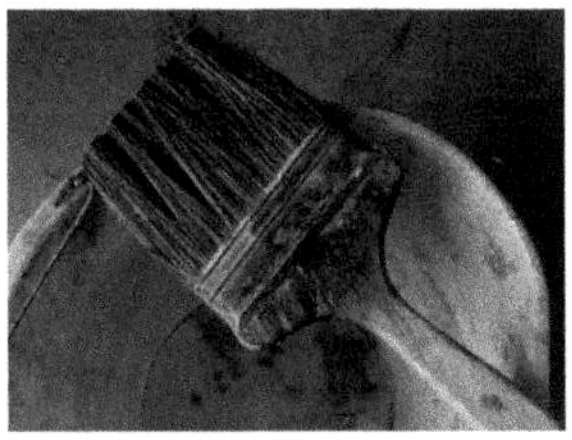	Used to apply water when finishing setting plaster.

Tool	Use
Small brushes 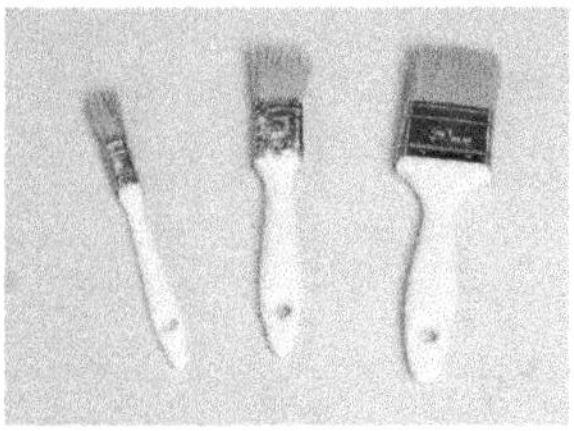	Used to clean internal angles and frames.
External corner trowel 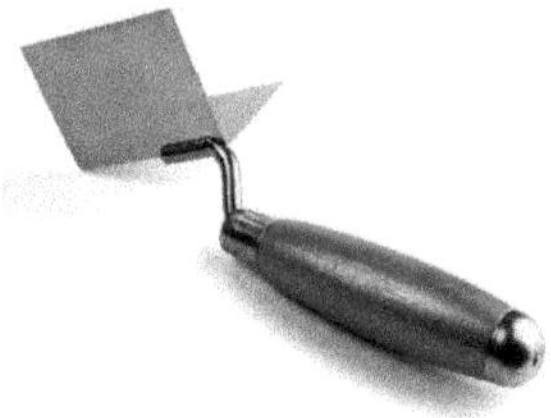	Used to form rounded hard angles, e.g. in walls with window openings.
Plumb bob 	A heavy weight attached to a string line, used to set out and transfer plumb points from above.
Internal angle trowel 	Used to form wet internal angles of finishing plaster during the setting and finishing process.

KEY TERM

Rule: flatten off plaster/render using an aluminium darby/straight edge rule.

Expanded metal lath (EML): sheet material in the form of diamond-shaped mesh that is used to reinforce a surface. This material can be fixed with screws and plugs or galvanised nails, or it can be bedded into the render material.

Key: referring to the background surface. A rough surface produces adequate key; smooth surfaces have less or no key.

Plastering materials, additives and beads used with internal plastering systems

You need to familiarise yourself with the materials and components needed to complete plastering activities. The following images show the majority of materials you will need when preparing and applying different plastering systems.

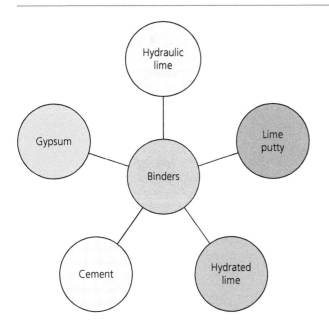

▲ Figure 3.1 Binders

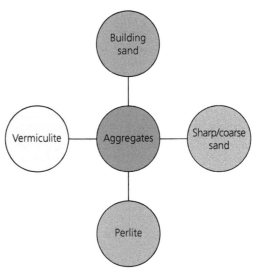

▲ Figure 3.4 Aggregates

▲ Figure 3.2 Cement

▲ Figure 3.5 Building sand

▲ Figure 3.3 Gypsum

▲ Figure 3.6 Sharp/coarse sand

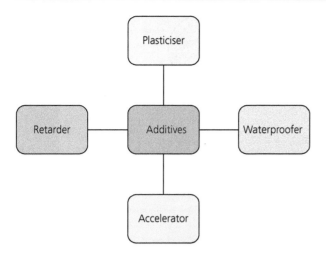

▲ Figure 3.7 Additives

▲ Figure 3.8 Plasticiser

▲ Figure 3.9 Waterproofer

▲ Figure 3.10 Accelerator

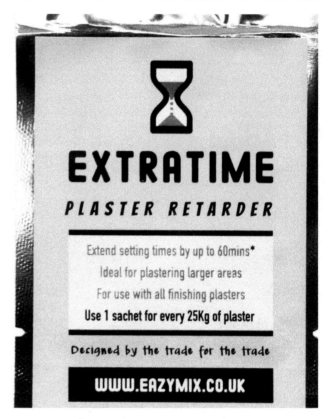

▲ Figure 3.11 Retarder

INDUSTRY TIP

Always ensure that you measure additives correctly. Adding too much or too little can affect the performance of the mix.

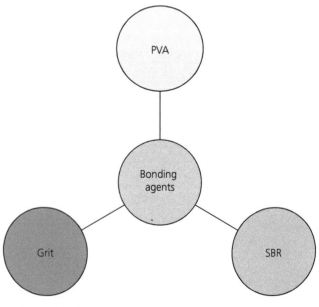

▲ Figure 3.12 Bonding agents

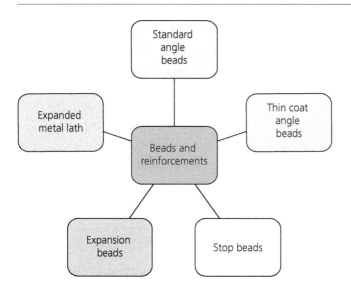

▲ Figure 3.13 Beads and reinforcements

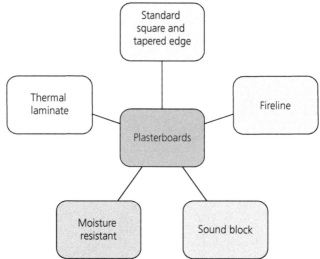

▲ Figure 3.16 Plasterboards

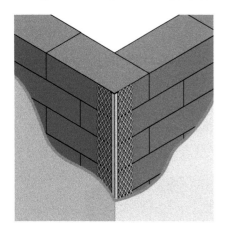

▲ Figure 3.14 Standard angle beads

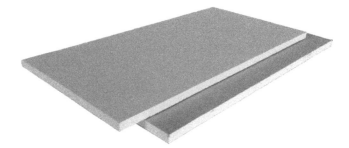

▲ Figure 3.17 Fireline

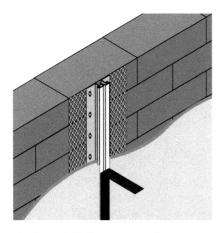

▲ Figure 3.15 Expansion beads

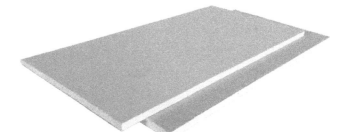

▲ Figure 3.18 Sound block

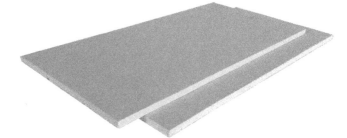

▲ Figure 3.19 Moisture-resistant plasterboard

▲ Figure 3.20 Thermal laminate

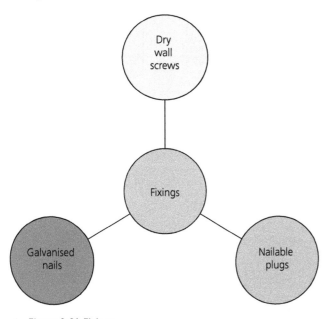

▲ Figure 3.21 Fixings

▲ Figure 3.22 Dry wall screws

Traditional backing coats

Traditional undercoat mixes used for applying pricking up, scratch and floating coats are referred to as 'lime mortar': a mix made of hydraulic lime putty as the **binder** and coarse sand as the **aggregate**, with the addition of horsehair for reinforcement. This helps to bind and strengthen the mortar and reduces **shrinkage**, which is a constant problem with this type of mix.

KEY TERMS

Binder: a material used to make the aggregate stick together when mixed.

Aggregate: a material made from fragments or particles loosely compacted together. It gives volume, stability and resistance to wear or erosion. Coarse- to medium-grained material used in construction for bulking.

Shrinkage: applied plaster can shrink as it dries out, forming small cracks.

Natural hydraulic lime comes in three different strengths. It has a chemical set that is similar to cement but far slower, requiring longer periods for drying and setting.

▲ Figure 3.23 Natural hydraulic lime

Hydrated lime relies on **carbonation**, so using this material in winter and summer can cause irregular drying.

▲ Figure 3.24 Hydrated lime

The setting coat is a mixture of lime putty with fine silica sand as filler or fine aggregate to bulk up the consistency.

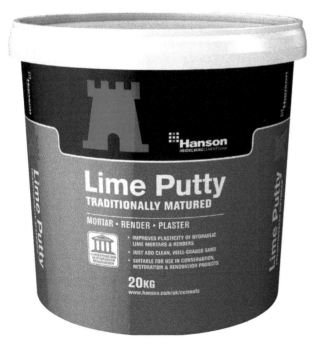

▲ Figure 3.25 Lime putty

Lime mixes were deemed weak and slow-setting; they could take several days, weeks or even months to set between applications, depending on the temperature. The use of lime could be a constant strain on plastering contracts, because deadlines might not be met due to the slow set.

Today, lime-based plaster mixes are still commonly used in conservation or **restoration work**. One advantage of lime plastering mixes is that they allow the walls to breathe, which reduces condensation. The plasterer will often begin by mixing and **batching lime mortar** the day before the work commences, and then remixing to increase it's workability.

▲ Figure 3.26 Conservation plastering using lime-based plaster

KEY TERMS

Carbonation: exposure to the air to start the setting process.
Restoration work: restoring plasterwork back to its original state.
Batching lime mortar: mixing mortar in preparation for the following day.

ACTIVITY

Research how lime and cement are manufactured.

Cement-based plaster mixes

Cement was introduced into the industry just after the end of the Second World War. Made of 75% limestone and 25% clay, cement is a binder that provides a faster set and also provides strength in the mix. The introduction of cement in undercoat mixes meant buildings and plastering work had faster setting times. This allowed work to be completed more quickly, which suited the high demand for new housing after the war.

Cement mixed with sand and water will begin to set after 45 minutes and is normally completely set by the next day. However, the mixture can take several days to reach its final strength. This process is known as **curing**.

Although cement had advantages over lime mixes, it also had its own problems, such as shrinkage, mixes being too dense or brittle and irregular strength resulting in cracking. These days:

- cement-based mixes are commonly used on **renovation work** or when harder background surfaces are required
- cement and sand plastering mixes are commonly used on old buildings that have been treated with a chemical damp course
- hydrated lime is sometimes added to cement and sand mixes to improve adhesion, workability and even increase **suction** for the finishing coat.

Cement- and lime-based plaster mixes should only be applied to solid backgrounds. The mix **ratio** commonly used is 5:1 sand and cement, or 6:1:1 sand, cement and hydrated lime. However, mix ratios can vary, to take into account the strength and compatibility of the background.

Lightweight backing and finishing plasters

Gypsum pre-mixed plasters have many advantages over traditional plastering materials because they have been designed and manufactured with a specific application in mind. During the manufacturing process, plasters are tested in laboratories to ensure they satisfy their intended use. The range of plaster produced today can be used on low and high suction backgrounds.

▲ Figure 3.27 Testing a background's suction by splashing it with water

Unlike cement- and lime-based plastering mixes, gypsum plasters contain manufacturer's instructions that provide user guidelines for each product. This provides additional performance benefits when specifying products for a particular design within a building. There are several different types of plaster available.

- Bonding-grade plaster is a pre-mixed undercoat plaster that contains vermiculite and perlite aggregate. Vermiculite is sharp and angular in shape, giving the plaster excellent bonding abilities compared with other types of plaster. This type of plaster is suitable for use on low suction and poorly keyed backgrounds, such as plasterboard or concrete.

▲ Figure 3.28 Vermiculite

- Undercoat plaster that contains perlite as the aggregate has to rely on the background having the necessary key to ensure the plaster bonds sufficiently. This is because the perlite aggregate is round and does not bond well to backgrounds that have a poor key.

▲ Figure 3.29 Perlite

- Thistle HardWall backing plaster is a pre-blended undercoat that contains additives to improve adhesion or cope better when there are high suction levels.

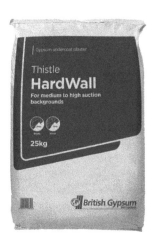

▲ Figure 3.30 Thistle HardWall backing plaster

- Multi-finish plaster can be applied to a range of gypsum undercoats and to sand and cement backgrounds that have been keyed with a devil float (see page 93).

▲ Figure 3.31 Multi-finish plaster

- Board finish plaster is applied to plasterboard backgrounds.

▲ Figure 3.32 Board finish plaster

- ThistlePro DuraFinish has been developed for use in areas that are prone to greater impact than normal, such as high traffic areas where people walking past may frequently knock into the plaster. Durable plaster provides 60% more protection against damage compared with ordinary gypsum finishing plaster.

▲ Figure 3.33 ThistlePro DuraFinish

- ThistlePro PureFinish is a finishing plaster designed to make indoor spaces healthier as it has Activ-Air® technology.

▲ Figure 3.34 ThistlePro PureFinish

ACTIVITY

Research the British Gypsum website to learn more about:

- backing and finishing plasters
- what type of backgrounds they can be applied to
- how much each bag covers.

Although gypsum plasters have advantages over traditional plaster mixes in terms of thermal values and heat resistance, gypsum undercoat plasters are not suited to older buildings which may contain areas that have been in contact with or affected by **rising** or **penetrating damp**. (This means the background might still contain some moisture.) Gypsum plasters have high absorption rates so this could result in problems if the dampness has not been treated and completely removed from the background. Gypsum plasters can become mouldy and eventually perish in persistently damp conditions.

▲ Figure 3.35 Damp wall affecting plaster

KEY TERMS

Rising damp: moisture rising up from the floor through the wall.

Penetrating damp: moisture travelling through the wall from outside.

Plastering manufacturers British Gypsum and Knauf have developed plasters that can be sprayed directly onto the background, saving time and labour by improving efficiency. However, spray machines can be impractical in residential properties due to the challenge of setting up the equipment in small, tight spaces.

IMPROVE YOUR MATHS

Visit the British Gypsum manufacturer's website and research Hardwall backing plaster and multi-finish plaster. For each type of plaster, work out:

- the coverage (the area covered by one bag)
- the number of bags required to plaster a room with a height of 2.4 m and a perimeter of 36 m.

2 PREPARE THE BACKGROUND FOR PLASTERING

Building methods and materials are always evolving, with many changes to the way we work and our practices. Modern materials have been developed and manufactured to improve the efficiency of mixing plaster and to ensure compatibility with different backgrounds, simplifying the plastering process. Despite this, plastering surfaces can still break down over time due to age, poor surface preparation or bad workmanship (including poor mixing).

The key to creating a good plaster surface is to identify and prepare the backgrounds beforehand. It is important to understand that not all backgrounds have the same properties. Some will be soft and weak, some will be hard and dense, while some backgrounds could be timber studwork, joists or metal **furrings**. Plasterboard fixed to old timber surfaces might need to be built out with backing plaster, depending on how uneven the surface is. This is quite common in old buildings that are being refurbished.

Different backgrounds need to be prepared for plastering in different ways. There are several steps you need to follow to make sure the plaster **adheres** well to the background surface. Before you start, you will need to control the suction and ensure the background has adequate key to ensure adhesion.

KEY TERMS

Furrings: metal stud wall or ceiling linings, fixed with plasterboard; also known as metal-framed backgrounds.
Adheres: sticks.

Suction tests

A suction test can be carried out by applying water on a solid background to see how much it absorbs.

Checking the suction will tell you if the background is dry and porous. You can do this by splashing water onto the background with a brush. The quicker the water is absorbed, the higher the background's suction.

- No suction or low suction (when the splashed water stays on the background's surface) will indicate that the background is hard or dense.

- High suction backgrounds (when the splashed water soaks in quickly) will absorb moisture from the plaster mix and might cause it to dry too quickly when applied.

▲ Figure 3.36 Metal and timber backgrounds

Types of background

This table shows the many different types of background that you might need to plaster and their requirements.

Type of background	Description and requirements
Hollow and solid blocks 	Newly constructed buildings that have block walling need little preparation before you apply plaster to their surface, because they have medium to adequate key. The surface is flat and can be plastered using traditional or modern pre-mixed plasters. Block walling built to today's specifications and standards needs only a backing coat and finish, known as 'float and set'.
Lightweight aerated blocks 	These blocks are lightweight and weak with an adequate key but high suction levels. Cement-based mixes are not compatible with these backgrounds because they are too strong for this surface. This type of block is best suited for pre-mixed plasters that are weaker than the background direct bond dry lining. Applying a solution of PVA (polyvinyl acetate – a water-based glue used for preparing background surfaces by improving adhesion) diluted with water (to the manufacturer's instructions) will seal the surface and control the suction.
Plasterboard 	There are several different types of plasterboard, but they are all made with a plaster core within an outer skin of paper. Plasterboard may have square or tapered edges. Before applying plaster, the boards need to be reinforced at their joints to prevent cracking. Plasterboard has a flat surface with low suction and only requires a finish coat, applied using a one coat system consisting of two passes of finishing plaster at an average thickness of 3 mm. If plasterboard is to be fixed over uneven timber studwork, **filling out** may be required; this should be done using a bonding-grade backing plaster that contains the aggregate vermiculite.
Timber lath 	Laths were traditionally used on timber backgrounds. They consist of thin strips of wood over which plaster can be spread. It can be time-consuming to prepare, fix and plaster this surface. This type of background is still used in the restoration of listed buildings.
Existing solid plaster 	This type of surface is common where solid walls require a 'makeover' due to poor surface condition that has developed over time. Over-skim makeovers enhance the appearance of a wall by re-skimming the surface without removing the old plaster from the background. When applying plaster to this type of surface, remember that it can only be as good as the background you plaster over. The surface must be solid and sound with no **hollowness**. Any flaking paint and surface grime or grease, that could prevent the new plaster from bonding, should be removed. This background has no key and, if it is painted, usually no suction, unless the background has different properties that are hidden behind the decorated wall surface. For example, if the wall has been replastered after installation of electrical services, it might contain different plasters on the background, creating different suction rates. A bonding adhesive should be applied to the surface of this background before it is replastered.

Type of background	Description and requirements
Existing plaster and plasterboard surface that has decayed over time 	There are many different types of plaster surface that might need to be replaced or restored, whether due to poor workmanship or deterioration over time. This type of surface might show signs of cracking, hollowness or a crumbling surface that cannot be decorated due to its condition. You will have to remove any existing surface finish before you are able to identify the background properties.
Clay bricks 	Clay bricks were very popular at one time and can be found in all types of building. A common fault with clay bricks is that they shell their face (the outer surface comes apart), causing the plaster to 'blow' (come away from the background). This type of background is often uneven because the bricks were manufactured in kilns at high temperatures, which made them all a slightly different shape. They were then laid on a lime mortar bed, which is very weak. Clay bricks and lime mortar joints have a high absorption rate that will cause high suction levels. This surface will need to be treated with a bonding adhesive before plastering. Raking out the joints will also improve the key.
Concrete common bricks 	These bricks are made from coarse aggregate mixed with cement. This surface is smooth and hard, which means the key is poor and the suction is minimal. A bonding **slurry** is best suited for this surface.
Concrete surface 	This is a hard, dense surface with poor key; it may have absorption and it needs to be **scabbled** if smooth. It will need to be prepared with a slurry if using sand and cement. However, lightweight bonding plaster adheres well with no need to prepare the background.
Engineering bricks 	This is a hard, dense surface with poor key and no absorption. The face of the brick has a glossy surface that is difficult to prepare for plastering. It has an enamel look and no suction. This surface needs to be scabbled to remove the sheen. You can then slurry the surface with a bonding adhesive. Alternatively, you can fix sheets of EML to the surface using mechanical fixings – this is a good way to reinforce and form a key on the background.
Stone and slate backgrounds 	These backgrounds are often found on very old rural buildings. Stone can have rough or smooth surfaces. Due to their irregular shape and size, stones create a very uneven background and require additional layers to build out the surface. The first of these layers is a **dubbing out** coat. Slate is similar to stone in that it can have uneven surfaces with a smooth face and no suction. The mortar joints between the slate and stone can be very thick and wide, creating large voids. The old mortar joints need to be raked out, filled in with a suitable plaster mix, then keyed with a comb scratcher. Any large stone or slate that has a smooth surface should be prepared with a bonding slurry to improve the bond.

Type of background	Description and requirements
Timber wall plates 	This background generally has low suction and no key. Timber can move and twist with moisture contact. Wall plates are used on top of walls as a fixing for roof trusses. EML is a good way of reinforcing the timber and providing a key.
Concrete lintels and pad stones 	This surface is generally flat and hard with minimal suction. Concrete is used to make lintels and pad stones, which provide load-bearing surfaces above openings such as windows. Again, if traditional sand and cement is to be used, this surface should be keyed and applied with a slurry unless a lightweight bonding plaster is used.
Composite backgrounds 	This means a background made up of two or more different materials. One method of preparing this type of background is to fix EML mechanically to the surface. This will strengthen and reinforce the background and the applied plaster.
Timber studs and metal framing 	Plasterboard is fixed to timber and metal using screws, but galvanised nails can be used when fixing to timber. Plasterboard can be finished with finishing plaster or tape and joint.

KEY TERMS

Filling out: building out an uneven background.

Hollowness: holes or depressions in previously plastered walls where the plaster has become loose from the background.

Slurry: a wet mix applied with a flat brush.

Scabbled: roughened.

Dubbing out: the application of several coats of plaster/render to achieve a greater thickness. Each coat is applied no more than 10 mm thick, allowing for setting between coats.

Removing old plasterwork from the background

Removing old plaster from backgrounds is a process known as hacking. It is important to remove all loose plaster from the surface. This can be carried out by hand or mechanically using various tools and equipment, but before you start you need to protect certain areas to prevent damage that can be caused by this type of work.

Plywood sheeting can be used to protect floors and openings such as windows and doors. Dust sheets and tarpaulins are good for protecting furniture that might be too heavy to move out of the building. Causing damage to the client's property is unprofessional and can be costly to repair!

▲ Figure 3.37 Hacking a surface

Tools and equipment used to prepare backgrounds

There are several types of hand tools that can be used to hack and remove old plaster. This type of work is strenuous and time-consuming, but it is very important that the tools are used appropriately to carry out the work efficiently.

▲ Figure 3.38 Lump hammer (top) and bolster (bottom)

▲ Figure 3.39 Pick hammer

▲ Figure 3.40 Scutch hammer

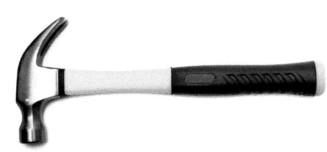

▲ Figure 3.41 Claw hammer

▲ Figure 3.42 Nail bar

▲ Figure 3.43 Wire brush

▲ Figure 3.44 Wheelbarrow

Power tools

Mechanical 'breakers' are far more efficient than hand tools and can remove hard plaster surfaces with ease.

▲ Figure 3.45 Mechanical breaker being used

Portable dust-extraction unit

Flying debris and high noise levels are some of the other hazards that you need to be aware of. Due to the dangers involved with this type of work, you will need to cordon off the work area with barriers to prevent any unauthorised person from accessing the area.

Tools for forming a key

Mechanical scabbling tools are used on dense smooth surfaces to create a rough area that will result in a good key, improving the bonding properties of the background. These tools have teeth (also known as pins) made from strong steel that vibrate, rotate and grind hard surfaces.

▲ Figure 3.46 Preparing a concrete background with a scabbler

Expanded metal lathing (EML)

EML can be mechanically fixed to a surface, especially if the surface contains large cracks or different composite. EML is also used on timber backgrounds, such as old timber lintels and wall plates.

▲ Figure 3.47 EML fixed to a background

Using bonding agents

Slurry coats are used to bond plaster to a background that has poor key but is still solid and strong enough to receive a plaster application.

▲ Figure 3.48 Wet slurry method

▲ Figure 3.49 Dry slurry method

Polyvinyl acetate (PVA) is usually mixed with water to seal and control suction in porous backgrounds. However, when used neat, it can bond finishing plaster to low suction backgrounds. When using undercoat plasters, it is advisable to mix the PVA with Ordinary Portland Cement (OPC) – this will then become a slurry and a strong **bonding agent** for bonding cement-based plasters to solid backgrounds.

KEY TERM

Bonding agent: a substance applied to improve adhesion on poorly keyed backgrounds.

Styrene-butadiene rubber compound (SBR) is another bonding agent, which can be mixed with cement to improve adhesion and waterproof surfaces affected by damp.

▲ Figure 3.51 Thistle Bond-it bonding agent

▲ Figure 3.50 SBR

Specifically manufactured pre-mixed grit adhesives are applied to solid pre-plastered and textured surfaces to form a key when over-skimming. They are available in various colours, depending on the manufacturer.

▲ Figure 3.52 Blue grit

IMPROVE YOUR MATHS

Research one of the manufacturers of grit adhesive and work out the coverage of a 10 litre container.

Preparing backgrounds for timber laths

▲ Figure 3.53 Timber laths

Timber laths were traditionally used to form a suitable background surface for plastering using three coats of lime plaster. Laths are thin strips of wood 1.5 m long, 30 mm wide and 6 mm thick. It is time-consuming to fix these strips to studwork and **ceiling joists** using nails. However, timber laths are still used today in conservation work.

Preparing backgrounds for installing plasterboard

Timber studwork, metal studwork, ceiling joists and **metal ceiling linings** are generally fixed with screws slightly penetrating the surface to avoid the trowel catching when finishing. Plasterboards are fixed to solid backgrounds with dry wall adhesive, which is known as **direct bond fixing**.

KEY TERMS

Ceiling joists: horizontal components of a ceiling, to which the rafters are attached.

Metal ceiling lining: a metal grid background to form accurate, level ceiling systems in old and modern buildings.

Direct bond fixing: using dry wall adhesive to fix plasterboards to solid backgrounds.

③ PREPARE TO FIX PLASTERBOARD

Installing plasterboard is the first task carried out by the plasterer, whether the walls are to be plastered using a solid plastering system or dry lined direct bond. The first task is to install plasterboard to ceiling joists, followed by fixing to the stud partitions.

ACTIVITY
Research how plasterboard is manufactured.

Characteristics of plasterboard

There are different manufacturing companies that make plasterboard but they all have one thing in common: they produce a range of plasterboard designed to meet and cope with the various requirements and demands of new homes built today. They also comply with building regulations and industry changes that often govern which type of plasterboard is used.

Plasterboard is manufactured in a range of sizes and thicknesses, each type with a specific design and use. The most common types used in the plastering industry come in two standard sizes:

- One is 2400 mm × 1200 mm and is designed to be used in new buildings that have walls of the same height, or studwork or joists at adequate centres. Using this standard size reduces waste from offcuts.
- The other plasterboard is slightly smaller and lighter at 1800 mm × 900 mm. This makes it easy to handle and fix if you are working on your own, or if the joist centres are uneven and vary in distance.

The plasterboard's edge type determines the type of finish surface it will receive:

- Square edge (SE) plasterboards are manufactured to be finished with setting coat plasterboard finish or multi-finish.

▲ Figure 3.54 Square edge plasterboard

- Tapered edge (TE) plasterboards are manufactured to be finished with a tape and jointing compound, which is sanded and sealed with a primer to protect the board's surface and prepare it for decorating.

▲ Figure 3.55 Tapered edge plasterboard

Planning for installation

Before you start to install plasterboard, you need to ensure that **first fix** electrical and plumbing services have been installed. These could include wiring sockets and lighting, telephone and media cables and pipes for hot and cold water and central heating. All of these must be installed before the plasterboard.

Different types of plasterboard are manufactured to meet the demands of industry and building regulations. Plasterboards are preferable to solid plaster as they have better performance levels within buildings, having the following properties:
- increased fire-proofing
- increased soundproofing
- repel moisture
- impact-resistant
- vapour barrier
- improved thermal values.

In older buildings requiring renovation, there may be many problems that need to be resolved before installing new plasterboard. Old timber joists will need to be **denailed** and it is likely that some of the timber will be rotten and need to be replaced and/or strengthened with additional **noggins**. Timber can also warp and twist over long periods and might have been fitted to different fixing centres than today's standard distances. A straight edge or string line should be used to check the **alignment of the joists**: if not corrected at this stage, the final surface may show steps and unevenness once the plasterboard has been installed.

KEY TERMS

First fix: all work (carpentry, electrical or plumbing) carried out before plaster is put on internal walls.

Denailing: removing old nails in timber stud and timber joist backgrounds before re-installation of plasterboard.

Noggin: a timber strut fixed between timber studwork or timber joists to strengthen and prevent twisting.

Alignment of the joists: checking the joist ceiling for straightness.

Tools required for preparing and installing plasterboard

Type of tool	Description
Utility knife	Used for cutting boards. Utility knives can have fixed blades but, for safety reasons, ones with retractable blades are better.
Tape measure	Essential for taking measurements from the backgrounds and transferring them accurately, allowing you to cut the board to the required size or shape.

Type of tool	Description
Straight edge or feather edge 	For providing a firm and straight guide against which to cut.
Hammer 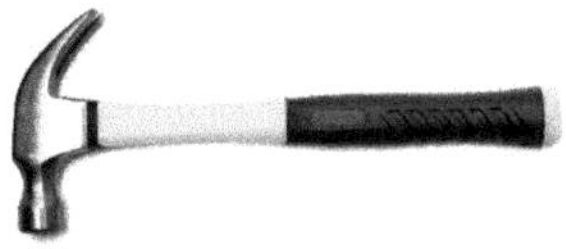	Used for fixing plasterboard by nailing. The broad head ensures that the nail's head is left below the surface of the board without piercing the paper face.
Drywall drill 	For fixing drywall screws, operated by battery or mains. Self-feeding screw drills are also available. A drywall drill has a built-in clutch that prevents the screws from being fixed too far into the board surface.
Rasp (rasp plane, plasterboard plane) 	Used for smoothing off cut edges or trimming down boards that are slightly too long. Curved cuts can also be formed using this tool.
Pad saw/service cutter 	Used for cutting out holes in plasterboards for electrical sockets or pipework.
Foot lifter 	Used to lift plasterboard up to the ceiling line.
Box rule 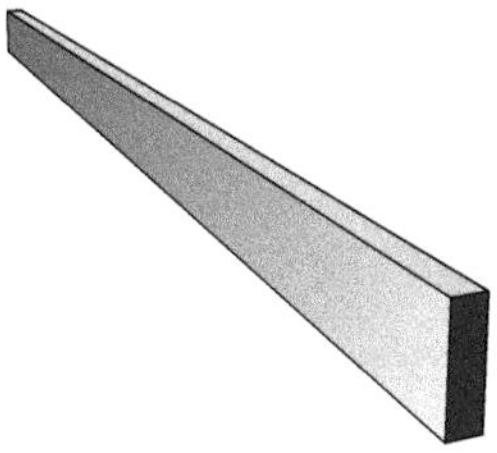	Used for installing and lining the plasterboard surface against the drywall **dabs**. It has a wider edge than a normal straight edge or feather edge rule, so it is less likely to cause impact damage to the board surface when tapped against the plasterboard.

Type of tool	Description
Chalk line	Snapped on floors and ceilings to form guidelines when installing wall boards.
Laser level	Used when setting out to provide accurate horizontal and vertical guidelines.
Spirit level	Used with a straight edge for plumbing and levelling plasterboard up to openings such as windows, doors and returns.

Storing and handling plasterboard

Plasterboards are fragile and easily damaged. They can **bow** if stored upright and if kept in damp conditions they will soak up moisture and **deteriorate**. Correct storage and handling are essential to help prevent damage from occurring.

- Plasterboards should be stored flat, under cover in dry conditions, off the ground and well supported.
- Because of the shape and size of plasterboards, two people are required to move them. A plasterboard should be carried on its edge, with a person at each end to support it. One person should lead the operation while the other follows instructions. Take care not to damage the ends of the boards when picking up and setting down, or when negotiating obstacles such as doorways and corridors.
- If plasterboards are carried flat, the core may crack, causing damage to the paper face and later becoming defective and cracking if they are fixed and plastered.
- When moving plasterboards outdoors, be aware of the weather conditions: a gust of wind could catch the board and blow you off balance. You must limit their exposure to rain and moisture and keep the stack well covered and protected.

KEY TERMS

Dabs: dry wall adhesive applied to the background to receive direct bond plasterboard installation.

Bow: bend.

Deteriorate: become damaged, defective and unusable.

▲ Figure 3.56 Moving plasterboard by hand

Procedure for fixing plasterboard with nails

Galvanised nails are used to fix plasterboard to timber studwork and ceiling joists, using a claw hammer or lath hammer. The nails are galvanised to prevent rusting.

When fixed, the head of the nail should penetrate the board, finishing flush with the plasterboard surface. The nail should penetrate the timber background by a minimum of 20 mm. Nails are fixed every 150 mm.

▲ Figure 3.57 Galvanised nails

Procedure for fixing plasterboard mechanically

Fixing plasterboards with screws means that the plasterboard surface is less likely to move under vibration. Another reason for choosing this procedure is the higher moisture content in timber studs or joists, which means they might move and twist as they dry out.

- Screws used to fix plasterboard to timber backgrounds must penetrate the timber by a minimum of 20 mm.
- Screws used to fix plasterboard to steel studs or linings must penetrate the steel stud or lining by 10 mm.

Drywall screws are considered stronger fixings than galvanised nails and have less chance of **popping** through the finished surface.

▲ Figure 3.58 Self-feeder fixing tools

Screws should be fixed without fracturing the board surface but firmly enough to penetrate the surface

for filling by **spotting** or finishing with setting plaster. Plasterboard screws have Phillips countersunk cross heads which are fixed with a PH2 drill bit.

Screw sizes can vary from 22 mm to 90 mm in length. They are fixed at intervals of a minimum of 300 mm to stud walls and every 230 mm into ceiling joists.

KEY TERMS

Galvanised: coated with zinc to prevent corrosion.

Popping: where plaster comes away from the plasterboard background because a fixing is loose or has been driven too far into the plasterboard surface.

Spotting: applying a small amount of jointing filler over penetrated fixings.

▲ Figure 3.59 Phillips countersunk cross head screw

IMPROVE YOUR MATHS

1 How many screws will you need to fix one sheet of plasterboard measuring 1200 mm × 2400 mm, fixed upright, if the studs are at 600 mm centres?
2 How many sheets of plasterboard will you need for a ceiling measuring 4.3 m × 5.2 m plus 10% for waste?

The following table shows the maximum joist centres for plasterboard, both with and without noggins.

Plasterboard thickness	Maximum joist centres with noggins	Maximum joist centres without noggins
12.5 mm	600 mm	450 mm
15 mm	600 mm	600 mm
19 mm	600 mm	600 mm
Thermal laminates	600 mm	450 mm

Procedure for fixing plasterboards to a ceiling

The following steps show the staggered method for installing plasterboard to a 10 m² timber joist ceiling. Before you start to fix plasterboard, set up a staging that allows you to install your plasterboard safely and efficiently to the ceiling.

1 Mark the wall to show the position of the ceiling joists.

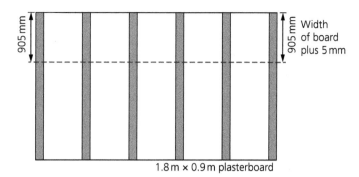

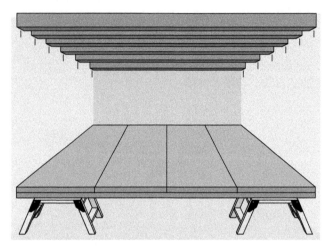

▲ Figure 3.60 Marked lines for the ceiling joists

2 Mark the width of the plasterboard, allowing an extra 5 mm on the joists at each end of the ceiling, and snap a chalk line. This will give you a guide line on the ceiling against which to install the plasterboard edge.

KEY TERMS

Plasterboard strut: used to prop the plasterboard in position prior to securing with screws.

Dead man prop: a telescopic pole with pads on each end. The pole is adjusted to hold an item above your head just like an extra pair of hands.

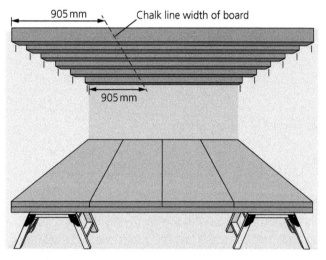

▲ Figure 3.61 Marked width of the plasterboard

3 Measure and cut the first board to the centre of the furthest joist.

4 Rasp the cut edge. This side will be positioned to the wall.

5 Using a **plasterboard strut** or a **dead man prop**, position the plasterboard along the chalk line, making sure the end of the board sits on the centre of the joist.

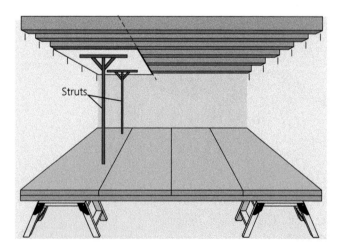

▲ Figure 3.62 Using struts

6 Once the board is correctly positioned, it can be fixed to the correct fixing centres using nails at 150 mm or screws at 230 mm.

7 Continue and fit the next board, leaving a small gap of about 2 mm after butting up to the previous board. Make sure you install the board to the chalk line.

8 On the adjacent run you will need to stagger the joints. You can do this by fixing a shorter plasterboard first, followed by a full plasterboard. You have now completed the layout of the staggered ceiling.

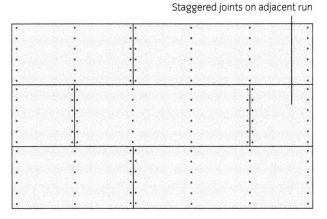

▲ Figure 3.63 Completed layout of staggered ceiling

▲ Figure 3.64 Horizontally plasterboarded wall

Procedure for fixing plasterboards to a timber stud wall

There are two methods that can be used when fixing plasterboard to studwork.

1 Fix the plasterboard horizontally across the studs. The adjacent run will need to be staggered, which will increase its strength and help reduce in-line cracks. This method is used more when using plasterboard of 1800 mm × 900 mm.

2 Install the plasterboard vertically, in line with the studs. This method is preferred on new buildings that have been designed to accommodate standard wallboards at 400 mm centres.

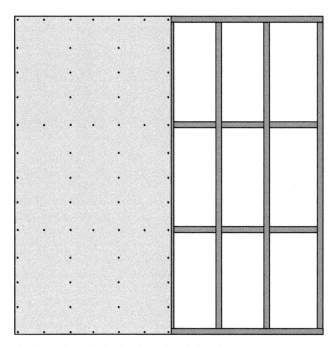

▲ Figure 3.65 Vertically plasterboarded wall

The following step-by-step instructions show the staggered method for fixing plasterboard horizontally to studwork that is 10 m².

STEP 1 Mark the floor to indicate the position of the studs.

STEP 2 Cut and position **packers** to avoid the plasterboard coming into contact with the floor.

STEP 3 Measure and cut the first board to the centre of the nearest stud.

STEP 4 Rasp the cut edge and position this side facing the internal wall, with the good edge to the centre of the stud.

STEP 5 Position the plasterboard, making sure the edge of the board fits to half the stud, and place a level along its edge before fixing with screws every 300 mm on studs, 230 mm on ceilings, or with galvanised nails every 150 mm.

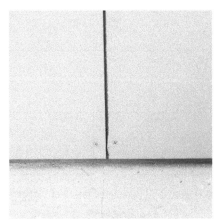

STEP 6 Continue to fix the boards along the wall, making sure there are no steps in the joints. Stagger the board on the next run and leave a gap of 2–3 mm at the joints, which will be filled with plaster and **scrim** to reinforce it.

STEP 7 The completed plasterboard **partition**.

KEY TERMS

Packers: small pieces of doubled-up offcut that the boards can sit on to keep them off the floor.

Scrim: used to reinforce plasterboard butted joints to reduce cracking before applying finishing plaster.

Partition: wall used to separate and divide the overall space within a building into rooms.

4 FIX PLASTERBOARD BY DIRECT BOND

Drywall adhesive is a pre-mixed product for direct bonding plasterboard to solid backgrounds. Fixing by direct bond is a preferred method because it is easy to install and has increased performance. It also reduces condensation, which is a constant problem with solid plastered walls.

The adhesive is mixed with water using a drill and whisk or a handheld mixing wheel, until it is the correct consistency to apply dabs of adhesive to the wall. A polymer additive is included in the adhesive to improve its adhesion properties, making it a suitable product for direct bonding plasterboard and laminate surfaces to the background.

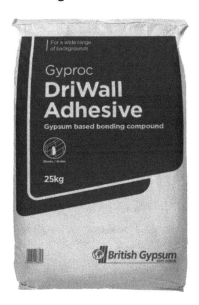

▲ Figure 3.66 Drywall adhesive

Suitable solid backgrounds for drywall adhesive include blockwork, brickwork, concrete or scratch coated (keyed) surfaces that may have been created to dub out uneven walls. There is a certain procedure for fixing plasterboards to solid walls and if this procedure is not followed precisely, the dry lining background will not perform to its design requirements.

- Avoid dry lining over damp or painted walls unless they can be treated and prepared beforehand, as plasterboard will perish and deteriorate in damp conditions.
- Check backgrounds with a straight edge to identify any high or low points in the wall, as this will determine how thickly or thinly you need to apply the dabs of adhesive.
- Before drywall adhesive can be applied as dabs to receive the boards, you need to position and line the dabs vertically and horizontally to the top and bottom of the wall. Mark the width of the board on the wall to indicate the line of dabs, which should be 600 mm apart. This will mean each full plasterboard has three vertical rows of adhesive. Apply a continuous line of adhesive at skirting level to provide a solid fixing for the skirting board.
- Before any plasterboard is installed, fill perimeter seals with adhesive to prevent air flow from entering behind the plasterboard. Air flow can also come from services, sockets and window and door linings that are fixed to external walls.

Setting out chalk lines

STEP 1 Measure 125 mm along the floor level from the wall at both ends. Transfer the marks up to the ceiling line at both ends of the wall.

STEP 2 Snap horizontal chalk lines on the ceiling and floor to form marked lines.

STEP 3 Mark the width of the plasterboard on the wall.

Procedure for fixing plasterboard by direct bond

STEP 1 Apply the dabs of adhesive to the wall in three vertical lines. Make sure you place the end dabs about 30 mm in from the edge of the plasterboard.

STEP 2 Apply a continuous line of adhesive to the base of the wall to create a solid fixing for the skirting and to the ceiling line for support.

STEP 3 Place two packers to the floor. Fit the board onto the wall and check that the edge of the board is plumb.

STEP 4 Use the box rule and tap the board against the dabs and wall until the edge of the box rule is in line with the chalk lines set at the floor and ceiling lines.

STEP 5 Lift the board to the ceiling using the foot lifter and pack underneath the plasterboard. Check the plasterboard surface with the box rule to make sure it is still plumb. Repeat the process for the next board until you have finished the wall.

Installing plasterboard around electrical boxes

If you are plasterboarding a wall which has electrical boxes fitted – for plug sockets, light switches or services, for example – then you need to cut gaps for these out of the plasterboard before fixing. The following step-by-step instructions show you how to cut out for electrical boxes.

STEP 1 On the wall, measure from where the bottom of the board is going to sit, up to the top and bottom of the box.

STEP 2 Transfer these measurements to each end of the plasterboard. Draw a line between these points to give you two parallel horizontal lines.

STEP 3 On the wall, from where one edge of the board is to be situated, measure the distance to each side of the box.

STEP 4 Transfer these measurements to each horizontal line on the plasterboard and draw lines between them to form the outline of the box.

STEP 5 Using a pad saw, carefully cut out the outline, taking care to keep to the lines.

Installing plasterboard around reveals

When you have walls with windows and/or door openings, it is best to fix the plasterboard edge 25 mm past the **reveal** or return. This allows you to fit a plasterboard sheet and dab in to the reveal or return. It is best to fix the head and sill first, as the reveal board can take the weight of the **soffit** board.

KEY TERMS

Reveal: small return to a window or door opening.
Soffit: the underside of a window or door opening.

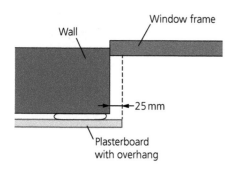

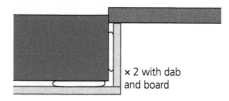

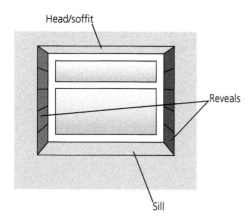

▲ Figure 3.67 Detail of overhang

1 Set out and snap chalk lines to both back walls, allowing a clearance of 25 mm for the thickness of dab and plasterboard. Using a square off the main walls, mark a right angle to both returns, then mark a line off the face of the pillar.

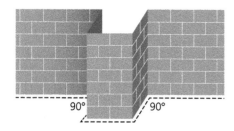

▲ Figure 3.68 Setting out on the floor off the main wall

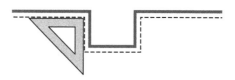

▲ Figure 3.69 Plan view of setting out returns off the main wall

2 Cut and fix the left- and right-side plasterboard, making sure you have the cut end to the wall and the factory edge to the set out mark at the ends of returns. This will also make it simpler to plumb the outer edge of the board.

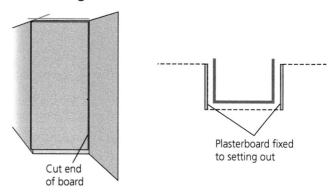

▲ Figure 3.70 Plasterboard attached to pier

3 Next, cut and fix the plasterboard that fits into the face of the pier between the two return ends and proceed to cut and fix the plasterboards for the main wall. This will cover your return cut ends at both internal angles of the returns.

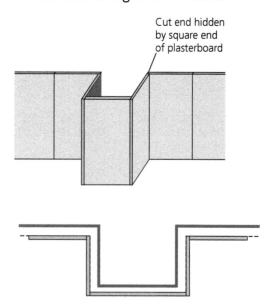

▲ Figure 3.71 Plasterboarded pier and main wall

Do not forget that if you are dry lining direct bond with thermal laminates, you will need to:
- allow for the thickness of the insulation when setting out your chalk lines
- fix two nailable plugs near the top of the plasterboard, 25 mm in from the edge. In the event of a fire in the building, the insulation will break down and melt with the heat; without the plugs, the plasterboard might fall away from the background wall as the insulation melts. This could make evacuation routes difficult to use and cause access difficulties for firefighters.

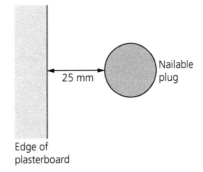

▲ Figure 3.72 Nailable plug 25 mm from plasterboard edge

⑤ PREPARE LOOSE PLASTERING MATERIALS FOR MIXING

Before setting up the mixing area, you must protect the floor area by placing tarpaulin or timber sheets on it to protect it from splashes and spillages that may occur when mixing. Dust sheets and cling film can be used to protect walkways and furniture.

It is extremely important to follow the specification when **gauging** plastering mixes and to gauge the materials accurately. Inaccurate mixing or gauging can lead to weakened mixes or mixes that are too strong.

KEY TERMS

Gauging: measuring out the ratio for mixing materials.

Gauging boxes were traditionally used to measure quantities of materials. Today, plastic buckets are preferred as they are lighter and have handles, which makes the job easier.

To measure quantities and volumes of materials accurately, fill the bucket or box to the top and then 'flatten off' with a piece of timber. This will ensure that the correct proportion of each material is added to make your mix.

Cement-based ratios for dubbing out and scratch coats can vary, depending on the characteristics of the background. In some cases, the background might have penetrating or rising damp, which is quite common in old housing. This will mean the ratio of the mix will be different compared to walls that have no damp issues but might have poor adhesion and key.

The following table shows the ratio of mixes for scratch coats and dubbing out.

Ratio of mixes	Use
Ratio (written as 6 : 1 : 1) of: 6 sand 1 cement 1 lime plus plasticiser **Ratio (7 : 1 : 1) of:** 7 sand 1 cement 1 lime plus plasticiser	These ratios would be preferred on severely uneven backgrounds, such as stone or old brickwork. In some cases, less lime is added; however, this will depend on the specification.
Ratio (3 : 1) of: 3 sand 1 cement plus waterproofer **Ratio (4 : 1) of:** 4 sand 1 cement plus waterproofer	These ratios would be preferred when dubbing out and applying scratch coats to backgrounds that have been treated for rising or penetrating damp.
Ratio (4 : 1) of: 4 sand 1 cement	This ratio is preferred when applying scratch coats to slurry surfaces.
Ratio (5 : 1) of: 5 sand 1 cement	This mix would be used on uneven surfaces that have good key and only require preparing with water or PVA.
Ratio (5 : 2) of: 5 sand 2 lime	This ratio is used for training purposes in colleges.

Ratios are used to show the proportions of a mixture.

You can increase the amount of each part, but you must do this in the same ratio. For example, if you need to order sugar and coffee in a ratio of 1 : 5, you need 1 kg of sugar for every 5 kg of coffee.

You could double (× 2) the amount of sugar and change 1 kg to 2 kg, but if you do, you should also double the amount of coffee, i.e. 5 kg should become 10 kg.

IMPROVE YOUR MATHS

Work out and calculate how many buckets of cement and lime you would need to mix with 48 buckets of sand, if the ratio is 6 : 1 : 1 (sand, cement and lime).

Mixing traditional cement-based plastering materials using machinery

Traditional sand, lime and cement materials are best mixed with a mechanical drum mixer, which will thoroughly mix the different materials that make the plaster mix.

- Clean water and any specified additive such as plasticiser should be added to the mixer first. This will prevent materials from sticking to the back of the drum.
- The consistency of the mixed material should be drier, rather than wet, when mixing is in progress. This will allow the additive time to make the mix workable and easier to use when applying the plaster.
- Mixing should be carried out for at least five minutes, allowing the materials to fully mix together.
- Do not forget to wear gloves/barrier cream, goggles/ glasses and boots to protect from splashes from the turning drum of the mixer.

▲ Figure 3.73 Mixing with a drum mixer

INDUSTRY TIP

Adding too much water will make the plaster slide down the wall, while too little water will make it difficult to spread.

Mixing by hand

There are some instances where smaller amounts of mix are required. This can be mixed loose on a flat surface or in a bucket. The following steps show how to mix cement-based plasters by hand.

STEP 1 First, gauge the materials.

STEP 2 Next, place the materials into a single pile.

STEP 3 Mix the materials dry (without adding water).

STEP 4 Once the materials are mixed, make a dip in the middle of the pile.

STEP 5 Measure the correct amount of plasticiser needed to improve workability. Add the plasticiser to a bucket of clean water.

STEP 6 Mix the plasticiser into the water.

STEP 7 Pour the water into the middle of the pile of materials.

STEP 8 Using a shovel, pull the dry material slowly towards the centre, into the water.

STEP 9 Mix and turn the material. The longer you turn, the better the mix will be.

STEP 10 The finished mix.

▲ Figure 3.74 Mixing with a plunger

INDUSTRY TIPS

Always read the manufacturer's instructions on the rear of the container when adding plasticiser to water.

Do not add too much water or the mix will become heavy, unworkable and difficult to use.

ACTIVITY

In groups of two or three, follow the procedure for mixing lime and sand mortar to a ratio of 3 : 1, plus plasticiser.

Pre-blended plasters

Pre-blended gypsum-based plasters can be mixed either by hand with a plunger or mechanically with a drill and whisk. When mixing, it is important to follow the manufacturer's technical instructions; these are normally printed on the back of the bag. Irregular setting can occur if you do not follow the rules of using clean water, tools and equipment when mixing this type of plaster.

Pre-blended plasters used to be mixed in baths with a rake or shovel, until a modern powerful motorised drill was developed with a whisk attachment. This tool can mix pre-blended plaster with ease.

▲ Figure 3.75 Mixing with a drill and whisk

⑥ APPLY PLASTERING SYSTEMS

The type of plastering carried out on internal walls within buildings will be determined by the background's surface. Internal plastering can be completed using several plastering application processes or systems.

Types of plastering application systems

- Plasterboard backgrounds require just one application, known as one coat.
- New blockwork requires two coat application. This is known as float and set.
- Severely uneven surfaces might require three coat application, called scratch, float and set.

Three coat work

Three coat work is generally applied on uneven backgrounds, building up the surface in three layers of plastering material to obtain the desired finish. Some severely uneven backgrounds might need more than one scratch coat to build out the surface.

Two coat work

Two coat work is applied on flatter backgrounds, such as block and brick surfaces, which can be completed using a floating and setting coat. This has an approximate overall thickness of 13 mm, but this might vary on uneven backgrounds. This surface does not require a scratch coat.

One coat work

One coat work is generally related to applying the finishing coat. This is carried out by applying two passes of finishing plaster to plasterboard or devil floated backing coats. The first pass is applied approximately 2–3 mm thick and left to **pull in** before the second pass is applied and trowelled up to the desired finish.

One coat work also relates to applying one coat of universal plaster approximately 10 mm thick and finishing in one process; this replaces the two coat process of float and set.

KEY TERM

Pull in: stiffen up or start to set.

▲ Figure 3.76 Floating coat

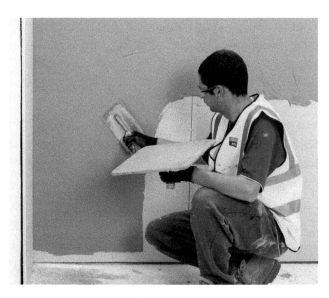
▲ Figure 3.77 Applying setting plaster to plasterboard

▲ Figure 3.78 Plastering a typical bedroom

▲ Figure 3.79 One coat plaster

Taping joints

Another way to finish plasterboard surfaces is to tape joints using pre-mixed or pre-blended joint adhesive, mixed by hand or mechanically. This procedure can be carried out manually or mechanically, using a range of accessories and materials.

The tape is available in roll form and has perforated holes and a centre crease in its design. These features make it easier to align the tape to straight joints and internal angles.

Corner paper tape is used on the external angles. It contains two corrosion-resistant metal strips along its length for strength and reinforcement, which means that this type of tape should be cut with a pair of tin snips.

Gyproc Habito Corner is a high-strength flexible corner tape that has a memory-free hinge. This means there is no need for pre-creasing or pre-measuring and the tape can fit any inside or outside corner angle.

Dry wall beads can also be used for forming and finishing external corners.

▲ Figure 3.80 Gyproc Habito Corner tape

▲ Figure 3.81 Paper tape

INDUSTRY TIPS

Avoid using dirty buckets – they can leave grit on the jointing surface that will be difficult to sand smooth and flush with the edge of the board joints.

Plasterboard and plaster manufacturers will only guarantee the joints against cracking if you use paper tape bedded in either plaster or jointing material.

▲ Figure 3.82 Corner paper tape

APPLYING THE SCRATCH COAT

Uneven surfaces need to be prepared by applying a scratch coat to the surface. This will build up and straighten its uneven surfaces, forming a base for the floating coat and controlling the suction. Once applied, the scratch coat is keyed with a comb scratcher to allow the next coat to bond. The scratch coat makes it simpler to apply and rule the next layer (the floating coat) with an even thickness of plaster.

The following steps show you how to apply a scratch coat. Before you begin, mix the different materials and include the plasticiser, which will give the mix its workability.

STEP 1 Load the spot board. Next, set up your hop-up. This will give you a platform to reach the ceiling height of the wall.

STEP 2 Wet the background surface, using a brush, to reduce the suction.

STEP 3 Transfer the plastering material from the spot board to the wall, using your hawk and trowel.

STEP 4 Apply the scratch coat, starting at the top right corner (if you are right-handed) or the top left corner (if you are left-handed), laying four trowel widths in lengths of 300 mm. Then flatten the plaster using the trowel at a shallow angle to help spread the material. The applied scratch coat should be about 10 mm thick.

STEP 5 Working downwards and across, follow the same procedure until you have applied an area of 1 m². It is best to break the wall surface into sections, as this is more efficient and will help you to complete the work in a methodical order.

STEP 6 This section of the surface is now ready to be keyed using a comb scratcher. Key the surface horizontally, deep enough to allow the next coat to grip but not so deep that you penetrate through the scratch coat to the background.

STEP 7 Apply the scratch coat to the rest of the wall and key it, using the same procedure.

STEP 8 Once the wall has been plastered, check its surface with a straight edge. Fill in any hollows or remove excessive thickness and prepare the surface to receive the next layer once the scratch coat has set.

STEP 9 Once the wall is complete, clean the work area: wet or messy surfaces are a slip hazard.

Applying the floating coat

Depending on the background and the specified plaster, the floating coat will provide a flat and lightly keyed base for the setting coat.

- Rule it with a straight edge, fill in hollows and remove any high spots in its surface that would cause unevenness after application.
- Then leave the plaster surface to pull in before consolidating and keying it with a devil float, preparing it for the final application known as the setting or finish coat.

▲ Figure 3.83 Devil floating

Lime and cement plaster mixes must be left to dry for several days before they can be finished with a setting plaster. Gypsum undercoat plasters set within two hours and should be finished with a setting plaster as soon as possible. If left for more than a day, gypsum undercoats can develop excessive and irregular suction rates that have to be treated with diluted PVA to prevent the surface from drawing too much moisture from the setting plaster, causing it to **craze crack**.

Once set, prepare all floated surfaces by scraping the surface with the edge of the trowel before applying the finishing coat. This will remove any **snots** or nodules which could protrude when the setting plaster is applied. Cut back all internal floated angles and ceiling lines to leave the surface flat.

KEY TERMS

Craze crack: when fine cracks appear on applied plaster, caused by excessive suction in background surfaces.

Snots: residue left on the surface of the floating coat after consolidation. This must be removed to prevent it from penetrating the surface of the setting coat.

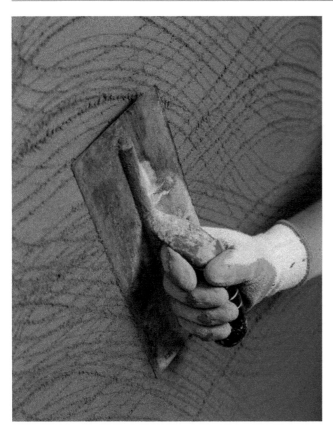

▲ Figure 3.84 Removing snots from the floating coat

▲ Figure 3.85 Cutting angles

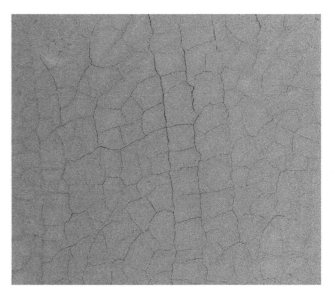

▲ Figure 3.86 Craze crack

Floating coats can be applied using the following three methods, each of which has its own purpose in terms of accuracy and speed of application:

● the plumb dot and screed method
● the broad screed method
● the free hand method.

Plumb dot and screed method

The plumb dot and screed process is the most accurate method of applying a floating coat, but also the most time-consuming. If the client is prepared to pay for this method, the end result is an accurate, plumb plastered wall.

Timber dots are used to set out accurate plumb walls and to form horizontal or vertical screeds (strips of plaster). The screeds are left to set before being used as guides, allowing the plasterer to fill between them and accurately rule the surface to obtain a flat plumb wall. The surface is later consolidated and lightly keyed using a devil float once the material has pulled in.

This system is sometimes used today in areas such as bathrooms or kitchens, due to the accuracy required to fit and fix units and sanitary tiles, or where fibrous plaster is to be installed.

INDUSTRY TIP

The plumb dot and screed method was often used in the early days of the industry.

STEP 1 Apply a dab of plaster approximately 150–300 mm from the ceiling and adjacent wall.

STEP 2 Set the dot into it. The distance from the wall surface to the face of the dot will be the thickness of the floating coat.

STEP 3 Directly below the top dot, set another dot approximately 300 mm from the floor.

STEP 4 Check that the dots are plumb.

STEP 5 When the dots have stiffened, apply plaster material between them to form the screeds.

STEP 6 Using the floating rule or feather edge, rule off the excess plaster between the dots. Remember to wear a hard hat when using a hop-up.

STEP 7 Fill any hollows. Repeat Steps 6 and 7 until the screed is flush with the dots and free from hollows.

STEP 8 When the screeds have stiffened sufficiently, consolidate the surface with a float and form key with a devil float. Remember to remove the dots and fill in with stiffening floating material.

STEP 9 Apply plaster between the screeds, starting at the top of the wall and working heel to toe.

STEP 10 Repeat until you reach the bottom screed.

STEP 11 Place the thinner edge of the rule across both screeds and, using a side-to-side motion, draw the edge up the screeds, ensuring that the edge remains in contact with both screeds at all times.

STEP 12 Any excess plaster will gather on the edge of the rule. This should be cleaned off the edge and returned to the spot board.

STEP 13 Fill any hollows until the plaster is flush with the screeds and all hollows are filled.

STEP 14 Repeat Steps 9 to 13 until the area between the screeds is completely covered and ruled off.

STEP 15 The finished floating coat.

Broad screed method

The broad screed method is also known as the box screed method. It also uses screeds as guides to help float the wall but is different from dot and screed.

Wet screeds are applied around the perimeter and ruled. Once the screeds have been laid on and ruled accurately, you must check that they are flush where they meet at the ends. The next stage is to fill in between the perimeter screeds and rule the surface either horizontally or vertically against the screeds. Although this method of floating coats produces straight walls and is more commonly used in the industry than dot and screed, the finished surface might be straight but not plumb.

When using this method, pay particular attention to all the internal angles. If these are not correct, it will be an obvious defect that will need to be corrected.

KEY TERM

Wet screed: band of undercoat plaster screed used as a floating guide while still wet.

STEP 1 Form the first screed up to 500 mm wide, applying plaster to the right-hand side of the wall from the ceiling to 25 mm short of the floor or damp proof course (DPC) and approximately at the required thickness.

STEP 2 Rule this off, checking it is plumb and filling hollows.

STEP 3 Apply a second screed along the ceiling line.

STEP 4 Rule in the ceiling line.

STEP 5 Apply another screed 25 mm short of the floor or DPC.

STEP 6 Apply the floating coat between the screeds and rule off. Take care not to scoop out the screeds as they will still be soft.

STEP 7 Ruling in.

Free hand method

The free hand method of producing floating coats is regarded as a way of completing the work efficiently and at speed, which is why it is best suited to large commercial sites where work is being done by subcontractors who have tighter work schedules than plasterers carrying out private work. However, due to the speed and the way it is produced, this method can be less accurate in terms of straightness. To help reduce this problem, this type of work is carried out on new block or brick backgrounds, which are much straighter than older uneven walls.

To carry out this work you need to be highly skilled and experienced, as there will be no guide such as screeds to help you rule your work. The plasterer will lay on plaster to cover a certain area of the wall before it is ruled using either a darby or a straight edge, filling in any hollows as the work proceeds. Ruling the surface is done horizontally and vertically to the applied surface to eliminate as much unevenness and to fill in as much as possible. Once the wall has been completely laid on and ruled, the surface will be checked at wall angles, ceiling lines, skirting lines and diagonally for straightness, before consolidating with a devil float.

INDUSTRY TIP

The industry standard for a plastered wall is ± 3 mm tolerance in a length of 1800 mm in any direction on the surface.

▲ Figure 3.88 Free hand method

▲ Figure 3.87 Checking free hand plaster with a straight edge

▲ Figure 3.89 Filling hollows

Floating ceilings and beams

Floating ceilings and beams involve a different technique from floating walls. Your tutor should be able to show you the methods below.

Floating a ceiling

Set out for floating ceilings using a bonding-grade undercoat plaster and then follow this method.

1 Mark a datum line around the room about 400 mm from the ceiling line.

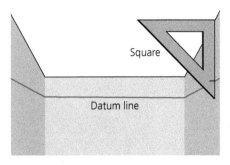

▲ Figure 3.90 Marking a datum line around the room

2 Using a tape measure, find the lowest point from the ceiling to the datum line; this will indicate the minimum thickness of your floating coat. At the lowest point close to the corner, start to set out your dots and apply your undercoat plaster to form a perimeter screed to the ceiling.

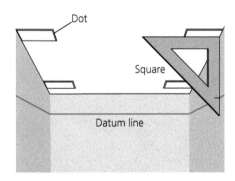

▲ Figure 3.91 Setting out the dots

3 Once the screeds have set and been keyed, fill between them with plaster, ruling off the screeds to form a level ceiling.

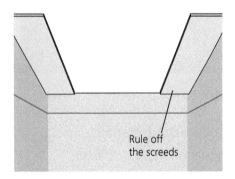

▲ Figure 3.92 Filled screeds and ruling off

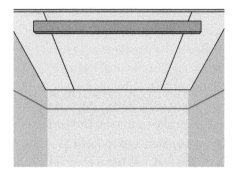

▲ Figure 3.93 Checking ceiling is level with a straight edge

If you have an uneven plasterboard ceiling with hollows, you will need to prepare and build this out by applying bonding plaster first, allowing it to set before applying the setting coat. Bonding plaster can be used to float plasterboard or EML that has been fixed to the ceilings.

Floating a beam

Follow this method to float a beam.

1 Fix angle beads and check for level and equal margin along the soffit of the beam.

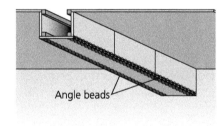

▲ Figure 3.94 Angle beads fixed to the beam

2 Apply the undercoat to the sides of the beam and then rule the surface with a square, ensuring it is 90° to the ceiling.

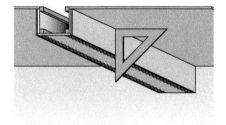

▲ Figure 3.95 Undercoat plaster applied and ruled on the sides of the beam with a square

3 Apply the undercoat plaster to the soffit of the beam and rule the surface of the plaster against the beads.

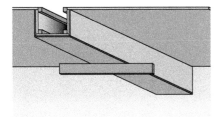

▲ Figure 3.96 Ruling the surface of the plaster against the beads

---

INDUSTRY TIPS

Make sure the angle beads are fixed level and the correct width gauge along their length. Use a piece of timber to mark the gauge, to check they are of equal length.

Beads are measured by length (linear measurement) and generally come in lengths of 2.4 m.

▲ Figure 3.97 Plastering a plasterboarded wall

IMPROVE YOUR MATHS

Calculate how many standard angle beads you need for fixing to four windows that each have two reveals and one soffit measuring 1.2 m.

Setting coat

The final coat will be applied in two thin layers of setting plaster using two different gauges. This coat is also known as the **skim**, which refers to skimming a surface with setting plaster. When this finish is completed, it will have an approximate thickness of 2–3 mm on walls and 3–5 mm on ceilings.

There are two types of surface in new housing that must be finished with setting plaster:
1 Plasterboard – a sheet material fixed to joists and studwork forming a base for the setting coat. This surface should be reinforced with a suitable scrim at its joints and where it meets the floating coat before the finish is laid on.
2 Ceilings – setting plaster is generally applied to ceilings before it is applied to adjacent walls, as this will reduce the risk of dirty water and plaster splashes affecting the surrounding surfaces.

Method for applying finishing plaster to a plasterboard ceiling (for a right-handed plasterer)

Make sure you have the necessary PPE to carry out this task. First, erect a fully boarded staging. This will allow you to apply your plaster in long strokes, making the job simpler, more efficient and safer to complete.

On some commercial sites, plasterers apply two applications of setting plaster using the same gauge. This can lead to imperfections such as sagging and **grinning** in the work. This can be avoided by using two different gauges.

KEY TERMS

Skim: the term used by some plasterers to describe the setting coat.

Grinning: when the plaster surface reveals imperfections caused by deeply keyed devil floating or variable suction of the background.

STEP 1 Apply scrim to reinforce the joints of the plasterboard, including the wall and ceiling line.

STEP 2 Apply plaster to cover the scrim. This will help keep it in place and prevent it folding when the first coat is applied.

STEP 3 Apply the first coat of plaster to the ceiling, working from left to right and laying the plaster in line with your right shoulder. This will help protect your face from any plaster droppings.

STEP 4 While plastering, make sure you stand in the correct position: put your right foot forwards and your left back. This will allow you a longer reach when applying the plaster.

STEP 5 Continue to work around the perimeter of the ceiling and then start at the opposite end.

STEP 6 The ceiling has now been completely covered with approximately 3 mm of plaster.

STEP 7 The next stage before applying the second coat is to flatten the ceiling surface with your trowel or finishing blade. Normally, the surface of the first coat will become matt in appearance, ready for the second coat.

STEP 8 Use a second gauge to apply your second coat in the same sequence as the first coat. Do not begin to apply the second coat until the first coat has started to pull in.

STEP 9 Once the second coat has been applied, flatten the surface, working in the same sequence as before. Make sure there are no blemishes or **galls** and that you have a good ceiling line along the wall.

STEP 10 Always lay and finish the plaster along the length of the bead to prevent any unforeseen steps or blemishes, which can be difficult to overcome at a later stage.

STEP 11 When the second coat begins to pull in, apply some water to the face of the plaster using a splash brush and trowel the ceiling, continuing to work from left to right in line with your right shoulder. Repeat this process every 15–20 minutes, at least twice.

STEP 12 Apply a final cross trowel to finish the ceiling to a flat smooth surface.

INDUSTRY TIP

When you lay on setting plaster to beads, you should always work your trowel along the length of the bead. Laying across the bead can create a recess in the plaster surface as it meets the edge of the bead.

Forming external angles

Walls with window openings

Before you plaster a window wall, you must prepare your angles. There are two methods for doing this:
1 Use **timber rules** as a guide, fixed or wedged to the reveals and soffit and positioned to allow for the appropriate thickness of plaster. This method is known as forming hard angles.
2 Use angle beads.

KEY TERMS

Galls: blemishes in a plaster surface due to poor workmanship.
Timber rule: straight plane timber used as a guide to form the edge of a return. Before the introduction of aluminium feather edges, timber rules were also used as straight edges.

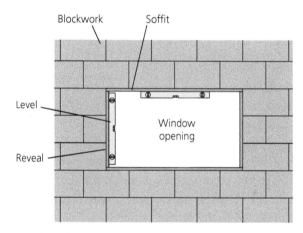

▲ Figure 3.98 Using timber rules in a window

Forming hard angles

Conservation work commonly specifies using this traditional method of forming corners. It can be time-consuming, because you can only apply plaster to one surface at a time and it needs to set hard before the timber rules can be removed and fixed onto the face of the wall, allowing you to plaster the reveals and heads. If you have to apply three coats, this will mean you have to fit and fix the timber rules three times, adjusting for each thickness of plaster application.

There are other types of external angle used to form hard angles, shown in the following table.

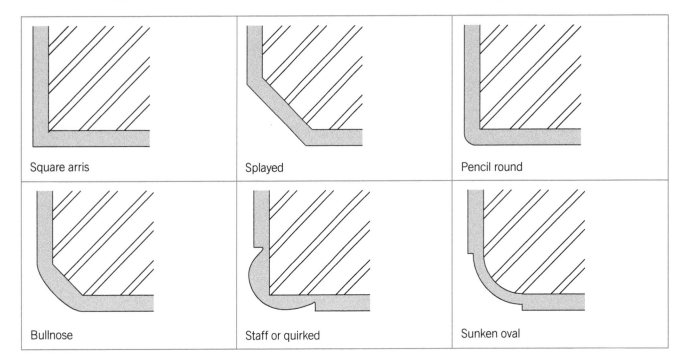

| Square arris | Splayed | Pencil round |
| Bullnose | Staff or quirked | Sunken oval |

The timber rule method is also used when applying the setting coat when you need to form rounded hard angles using an external corner trowel. The setting plaster is applied to the external corner of the floating coat in 50 mm wide strips each side. This is then trowelled with the corner trowel several times, to form and finish the corner as it sets. Setting plaster is then laid and finished to the corner's edge.

Reveal gauges are used to rule the floating coat on the reveal and soffit, providing a square return surface with equal margins along the frame.

Fixing angle beads

Standard angle beads are used not only to speed up the work but also because they provide accurate sharp corners that are reinforced to protect against impact damage, helping to prevent chipping of the external angle.

▲ Figure 3.99 Window fitted with angle beads

▲ Figure 3.100 Reveal gauge

▲ Figure 3.101 Using plaster dabs to fix angle beads

If severely uneven walls are to receive three coat work, timber rules or a straight edge might be used on the reveals to build out the unevenness first, before applying the angle beads.

The procedure for setting out and fixing angle beads is similar for a wall with doorways and returns. Standard angle beads can be fixed with plaster dabs or galvanised nails. The preferred method is to use plaster dabs when fixing, as this is simpler when plumbing or levelling the bead. Positioning beads with nails and plumbing at the same time can be a difficult process to master.

INDUSTRY TIP

When you fix beads with plaster dabs, remember to cut back any excessive dabs (that might stick out past the nosing) before the plaster sets.

The following step-by-step instructions show how to fix angle beads to a window wall. This method will allow you to set out the beads to the required depth and dimensions.

STEP 1 Place the straight edge 10 mm off the line of the wall, then use a pencil to place a mark along its edge. This mark on the sill indicates the thickness of the undercoat plaster that will be applied to the wall. If there is more than one window, a chalk line can be snapped from the outer side of each opening to indicate the line of the horizontal beads.

STEP 2 Use a square to mark a right angle from the window frame to the left and right reveal, making sure this is a similar thickness. Check that you have an equal margin along the frame's edge, as this will be the finished edge of the plaster at the reveals.

STEP 3 The marks have now crossed on the sill, indicating the fixing points for the two angle beads.

STEP 4 The next stage is to cut both angle beads just short (by about 10 mm) of the head soffit. This is to allow for the remaining bead that fits to the head; this can be cut after both sides have been fitted with beads.

STEP 5 Mix your plaster stiffer than normal and apply plaster dabs to the reveal. Pinch your plaster away from the corner, leaving a dab. Apply dabs every 400 mm (this can be less if the window is small).

STEP 6 Place the angle bead to your crossed lines at the sill and bed it in position, ensuring it lines up with the edge of the window. Remove excess plaster from the dabs once the bead is in position.

STEP 7 Use a level to check that both sides of the bead are plumb in line with the face of the wall and reveal.

STEP 8 Cut your head bead and repeat the process for setting out and fixing. This time make sure the bead is level along the head and that it does not have steps at either side to the reveal beads.

STEP 9 Allow the plaster dabs to set before applying the undercoat, as the angle bead will be used as a guide to rule the surface.

Use a square or gauge to form the reveals and head. The undercoat will require **cutting back** at the edge of the bead, allowing the setting plaster to finish flush with the **arris**.

▲ Figure 3.102 Using a square or gauge to form the reveals and head

▲ Figure 3.103 Cutting back the undercoat at the edge of the bead

There are other surfaces that have returns and angles. These will also require setting out before the angles are formed, using the hard angle method with timber rules or fixing angle beads in plaster dabs.

Walls with door openings

Walls that contain timber door casings or **linings** are very helpful to the plasterer as the lining can be used as a guide to rule the surface. However, if the lining is not fixed plumb and true, this will cause the floating coat to be a similar shape to the frame. Always check the door linings to make sure they are plumb and that they are fixed to allow a **nominal** thickness of plaster to the wall surface. If you find linings or frames to be out of line or not level or plumb, report the discrepancy to your supervisor as soon as possible, so the defective work can be rectified.

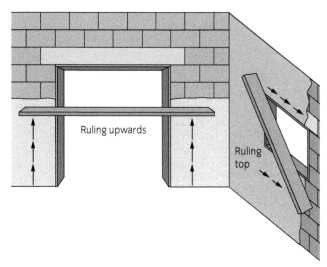

▲ Figure 3.104 Ruling a door and window lining

KEY TERMS

Linings: timber surround for internal doors, forming a lining to the masonry or studded opening.
Nominal: standard.

INDUSTRY TIPS

Hardwood door linings should be protected from wet plaster by taping with waterproof masking tape. Wet plaster can cause staining to the timber that cannot be removed.

Always clean door frames with clean water at regular intervals to prevent the build-up of plaster.

If the wall contains a return at its end, a standard angle bead must be fixed to the return. This should be lined through from the door lining to the corner, making a mark on the floor to indicate the thickness of the floating coat and the position of the bead.

Another method is to fix a timber ledge or rule to the face of the return to the same line as the door lining; this is later removed and reversed to complete the other side of the return.

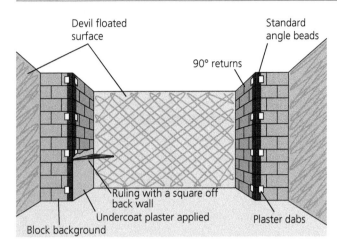

▲ Figure 3.105 Standard angle bead fixed to a wall with a return and forming returns with a square

After consolidating the plaster surface, you will need to cut back to the edge of the frame; this will allow you to apply a setting coat of plaster flush against the frame. Failure to do this can cause excessive build-up of plaster on the frame.

Walls with attached piers

For walls with attached piers, such as pillars, you need to set out snapping chalk lines at each end of the pier at floor level to the thickness of the undercoat. Once you have snapped your line, use this as a plumbing point for fixing the angle beads or timber rule. If you fix to the snapped lines, this should ensure that all beads or timber rules line up.

You can also use this method when setting out independent piers in line with each other. The internal right angle to the face and rear of the pier can be set out using a square from the snapped chalk line which runs along the length of the pier.

Walls with a return

Walls with a return can be set out using a square similar to a reveal; the square must be positioned at 90° from the line of the main wall. Figure 3.110 shows the floor marked out so that a wall with a return can receive standard angle beads.

▲ Figure 3.106 Snapped chalk line

Walls with a chimney

Walls with a chimney have two returns at each end, similar to a pillar. Once the beads are set out, you can check the margin from each side of the bead to the other at the top, middle and bottom for accuracy.

Walls with an independent pier

When setting out an independent pier with equal sides for fixing beads, you must set out a parallel line to the front and rear of the pier to the required thickness (approximately 10–20 mm). The next step is to use a square, set it to the parallel lines and mark the same thickness to both sides, making sure the sides are at 90°. Once you have set out your marks, you can fix your beads plumb to form the external corners. For accuracy, check the beads for equal margins at the top, middle and bottom.

▲ Figure 3.107 A wall with attached piers

▲ Figure 3.108 Wall with a return

▲ Figure 3.109 Wall with a chimney

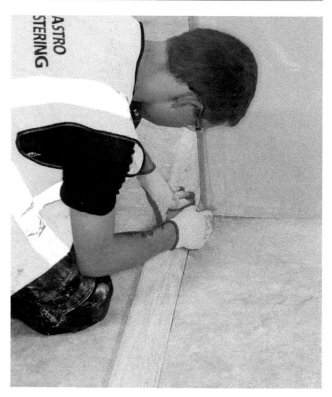

▲ Figure 3.110 Setting out an independent pier with equal sides

Beads

Different types of beads

Angle beads

Standard two coat angle beads are made to be used on two coat work. When fixed, they allow for a suitable plaster thickness, including the floating and setting coat.

They can be fixed using several methods, but the majority of standard angle beads are fixed with dabs of stiff setting plaster. Setting plaster is preferred because it is a fine material that allows the mesh wing of the bead to bed into the plaster with ease. We might also use undercoat plaster to bed beads, but in this case it is best practice to apply a continuous dab along the full length of the bead.

Thin coat angle beads can be fixed to solid or plasterboard corners. They are a smaller type of angle bead that allow for the thickness of setting plasters. There are two types of thin coat bead: mini mesh beads and solid beads. Thin coat beads can be fixed with galvanised nails, drywall screws, dabs of plaster or staples.

▲ Figure 3.111 Mini mesh beads

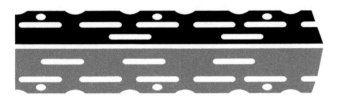

▲ Figure 3.112 Solid beads

Stop beads

Stop beads are used when the plasterwork needs to stop at a different surface. This could be when it goes up to timber or decorative masonry. Stop beads can be fixed with galvanised nails or bedded into position, depending on how **true** the background is.

Thin coat stop beads are used in the same way as standard stop beads. These beads are very popular when forming splayed angles that are not at right angles.

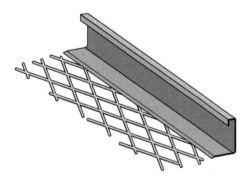

▲ Figure 3.113 Stop beads

KEY TERM

True: accurate to plumb, level and/or line.

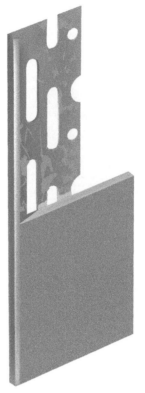

▲ Figure 3.114 Thin coat stop beads

Expansion beads

Expansion beads are used to prevent cracking on walls that have expansion joints to allow slight movement in both sides of the wall. These areas are considered weak, so expansion beads are fixed in line with the edge of the joint, allowing the plastered wall to move freely without cracking.

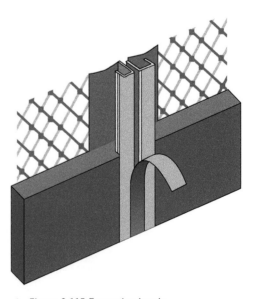

▲ Figure 3.115 Expansion beads

How to store beads

Store galvanised beads on a flat surface or upright in racks, away from impact or pedestrian traffic routes, to avoid any possibility of damage.

Beads are protected with a galvanised coating to prevent rusting when they come in to contact with wet plaster; however, if left outside in wet conditions for long periods they can corrode and become unusable. Do not use corroded beads: if used, they will rust and cause a pattern stain along the length of the bead that will penetrate through to the plaster surface. This can be difficult to remove.

Tape and jointing

Before you tape and joint plasterboard, you need to check that the plasterboard fixings have been driven home into the timber or steel, allowing you to fill the heads of the fixings. You should also check that there are no steps in the edges of the plasterboard and that any large gaps in the plasterboard joints are filled.

Tape and jointing is a process that produces a smooth, crack-resistant surface to the joints where TE (tapered edge) plasterboards meet. The screw or nail head fixings can be filled over with jointing material and sanded for priming and decorating. The jointing material should be left to dry before sanding and priming.

- You need to apply your jointing material first, before applying a paper tape over the joint to reinforce it. Before applying a second coat, squeeze and flatten the tape, allowing the next jointing application coat to cover the tape's surface.
- Before applying another layer of jointing compound, you must allow the previous surface to dry sufficiently as this will make it easier to cover. Some jointing compounds dry and set within one hour, while others are made from air-drying material that needs to set overnight.

- You can decide whether to use ready-mixed or pre-mixed jointing materials, as the same rules apply when applying and finishing. Make sure the material is feathered out and finished to the edge of the board.

▲ Figure 3.116 Sanding the jointed surface

▲ Figure 3.117 An internal angle

Tape and jointing angles

Internal angles can be finished by folding the tape in half and applying to adjacent walls. Take care when finishing opposite walls, as the trowel can dig in and damage the surface. A good way to avoid this is to use a corner tool, which is designed to finish both internal angles at the same time.

External angles can be finished by using either a drywall bead or a reinforced paper tape, depending on the type of reinforcement required. Both are bedded into the jointing material and finished in the same way as paper tape jointing. Once the surface has been jointed, it can then be sanded.

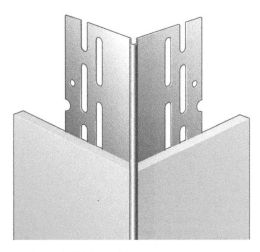

▲ Figure 3.118 Drywall bead

▲ Figure 3.119 Reinforced paper tape being used on an external corner

Tape and jointing method

The following step-by-step instructions show the method for tape and jointing standard TE plasterboard joints.

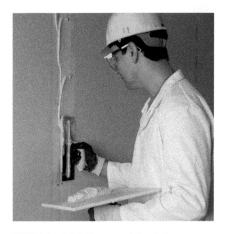

STEP 1 Apply jointing material and the paper tape to the surface, firmly squeezing the tape against the jointing. Make sure you cover the surface and the small perforated holes in the tape.

STEP 2 Apply more jointing material over the tape, making sure you feather out each application in turn, slightly wider than the first. For example, if the first application is 20 mm wide then the second application should be 30 mm wide.

STEP 3 Sand the surface to remove any unevenness and imperfections in the jointed surface. Using a roller and brush (as shown), apply a sealer or primer to the plasterboard surface.

Taping should be carried out carefully. The following table shows the faults that can occur if taping is not performed correctly.

Fault	Problem
Oversanding	The tape will show through.
No filler behind the tape	Will cause the tape to bubble (no adhesive behind it).
Uneven surface	Will cause the tape to be prouder than the board surface.
Fixings that have not penetrated the surface of the board	Will cause bumps and unevenness.

HEALTH AND SAFETY

When you carry out tape and jointing, you need to consider your own safety and that of other operatives who could come into contact with the dusty atmosphere created by sanding the joints and fixing surfaces.

Apply the following safety measures:
- Wear appropriate safety clothing to protect yourself from the dusty atmosphere.
- Electrical sanding tools should contain dust bags (or the dust should be extracted with an appropriate extractor) to reduce increased or high levels of dust that can be hazardous to lungs.
- Due to the amount of dust in the atmosphere, adequate environmental protection should be carried out before sanding. Use dust sheets or cling film to protect furniture, flooring and surroundings.
- Display warning signs to let others know about the dangers and hazards involved with the work in progress. Use barriers to keep people away from the work area.

▲ Figure 3.120 Warning sign

▲ Figure 3.121 Dry lining dust extraction

Parge coats

A parge coat, such as Thistle SoundCoat Plus, is a pre-blended, bagged material mixed with water and applied to party wall blocks before the installation of direct bond dry lining. This layer reduces noise transfer through the walls.

▲ Figure 3.122 Parge coat

When dry lining interiors, you must meet building regulations standards. These include requirements relating to resistance to the passage of sound and acoustic standards for separating partition walls and ceiling/floor constructions (Part E) and the redraft of Part L (relating to conservation of fuel and power) that raises the bar for energy performance.

Plasterboard manufacturers have developed new systems to comply with the Part E and Part L regulations. For instance, manufacturer British Gypsum has developed a pre-mixed soundproofing plaster parge coat that minimises air leakage to improve the thermal and acoustic performance of masonry walls, such as **party walls** that divide semi-detached and terraced dwellings.

Sealing materials

Drywall sealer is another product that is applied after taping and jointing to control vapour in bathrooms and kitchens. Otherwise, the large amount of moisture that exists in these rooms will penetrate the plasterboard surface.

Other sealing materials include special fixing and sealing sealants and mastics that are used with fire and acoustic performance dry lining systems. You can learn more about these products if you search the websites of manufacturers such as British Gypsum, Knauf and Siniat.

Sound and air tests

Buildings that contain dry lining surface linings will be tested for air and sound transmission.

Air tests are carried out on modern new builds. This means that all perimeter surfaces must be sealed, including internal angles to walls, ceilings and floors, frames to openings, and services. All possible air passage points into the building must be blocked.

Sound tests are carried out to check the acoustic performance of the building, identifying the need to reduce or eradicate sound that can have a severe effect on people's privacy in multiple-occupancy buildings.

Making good a chase

You might have to make good a **chase** if plasterwork has become defective and been replaced, or if new services have to be installed behind the original plasterwork. Once the work has been carried out, the chase has to be made good to return the plasterwork to its original state.

The following steps show how to make good a chase. First, you must mark out the chase on a plastered wall. Using a lump hammer and bolster, remove the plaster to leave a chase in the plastered wall, then follow the steps below.

KEY TERMS

Party wall: a dividing partition between two adjoining buildings that is shared by the occupants of each residence or business.

Chase: a void for installing services which will need to be made good with plaster.

INDUSTRY TIP

It is important to wear safety clothing and glasses to avoid flying debris when cutting out a chase.

STEP 1 Remove the dust and dampen the background to control the suction.

STEP 2 Mix the undercoat plaster and apply directly into the chase.

STEP 3 Remove excess plaster with your trowel in line with the existing surface, filling out any hollows.

STEP 4 Clean off the previously plastered face, using a trowel and a water brush.

STEP 5 As the undercoat plaster starts to pull in, cut back the surface with a gauging trowel or small tool to form a recess of approximately 2–3 mm in the chase surface to allow for the setting coat.

STEP 6 Key the undercoat with a devil float.

STEP 7 Soak the edge around the chase several times to control the suction at its edge before you apply your setting coat.

STEP 8 Mix your plaster and apply it to the chase, removing any surplus plaster that is applied beyond the edge of the chase.

STEP 9 Remove any surplus plaster away from the edge and clean the surface with a splash brush.

STEP 10 Repeat the process and apply the second coat.

STEP 11 Trowel the face of the plaster chase, making sure the surface is flush with the edge.

When you have completed this task, run the palm of your hand along the surface to feel if there are any recesses or bumps between the chase and the previously plastered face.

CLASSROOM ACTIVITY

After reading this chapter, work in groups to discuss and plan how to plaster a modern two-bedroom house built of standard blockwork containing timber partitions and ceiling joists. Decide on a two coat lightweight plastering system, performance plasterboard, fixings and different types of beads.

Workshop activity 1

Working with a partner, set out a scaffold staging in a bay and fix plasterboards to a ceiling. Ensure the boards stagger on the second run, then reinforce with self-adhesive scrim and apply finishing plaster in preparation for decoration.

Workshop activity 2

Direct bond around a pillar with plasterboard, fix thin coat angle beads to the corners and apply two passes of finishing plaster.

Workshop activity 3

Fix standard angle beads to a window wall, apply lightweight backing plaster to the wall face including the window returns and soffit, cut back at beads and devil float, then apply two passes of finishing plaster.

CASE STUDY

Megan's first house is a mid-1850s Victorian terrace, which she has purchased in the city. She preferred to purchase an older property rather than a newly built house. Read her story to learn how she got on.

I recently purchased a Victorian house that required some renovation work to modernise and upgrade it. When I bought the house, several plastering defects were identified in the surveyor's report. The plaster on the ceiling is loose and there are widespread cracks on the surface. The living room wall has a large deep crack running from the ceiling line down to the floor. I am eager to get these defects corrected and have been given the name and telephone number of a qualified plasterer to contact. He has been highly recommended after completing some renovation work for a family member.

The plasterer recently called at my property to give me advice on replacing and renewing the ceilings and making good the defective wall surface.

He has stipulated in his estimate that the rooms will need to be clear of furniture with adequate protection and a portable ventilation appliance should be hired to remove hazardous dust when taking down the lath and plaster ceilings. He suggests replacing the old ceiling with fireline plasterboard to increase the fire rating performance in line with building regulations, because fireline plasterboard provides 60 minutes of fire resistance. This should then be plastered with two passes of Thistle Board Finish 2–3 mm thick, polished flat and smooth ready for redecoration.

After carrying out a thorough check on the deep crack in the wall, his advice is to remove all old plasterwork from the wall surface, mix coarse sand and cement, fill the crevice flush with the face of the wall and then leave it to set hard. Once set, he has suggested fixing EML over the cracked area with mechanical fixings, as this will reinforce the weak area. The wall should then be plastered using a lightweight plaster system and this will require three coats: scratch, float and set.

As the plasterer's advice seems knowledgeable and precise, I'm going to ask him for a written quotation for the work. This will ensure that the work he carries out will be completed with a guarantee and to the required industry standards.

Megan's advice

It is important to find someone reliable and reputable to advise and carry out renovation work, as repairs and remedial work can be costly.

Test your knowledge

1 What aggregate is added during manufacture to lightweight backing plaster to improve adhesion?

A vermiculite

B perlite

C cement

D lime

2 What term is given to measuring loose materials when mixing plaster backing coats?

A batching

B gauging

C adding

D raking

3 What is the term given to the hardening process of cement and sand mixed for plastering?

A drying

B setting

C curing

D shrinkage

4 What type of defect occurs when finishing plaster is applied over a high suction background?

A grinning

B shrinkage

C blistering

D crazing

5 What reinforcement is fixed over timber wall plates before plastering?

A expanded metal lath

B self-adhesive scrim

C scrim cloth

D mesh matting

6 What type of plasterboard is best used on bathroom walls?

A fireline

B vapour check

C moisture-resistant

D sound block

7 When fixing plasterboard to block background, which procedure should be carried out?

A direct bond with drywall adhesive

B mechanically fixed with screws

C using a dry wall self-feeder

D fixing with galvanised nails

8 When fixing plasterboard to ceilings, what are the recommended fixing centres of the screws?

A 260 mm

B 300 mm

C 230 mm

D 350 mm

9 What is the recommended thickness when applying finishing plaster to floated backgrounds?

A 8–9 mm

B 6–7 mm

C 4–5 mm

D 2–3 mm

10 After ruling a floating coat, what is the next process before applying the finishing coat?

A devil float the surface

B key with a comb scratcher

C fix angle beads with plaster

D apply scrim to the joints

11 What type of bead is used on returns built of blockwork background?

A stop bead

B standard angle bead

C expansion bead

D mini mesh bead

12 When is the process of cutting back carried out?

A after devil floating

B after applying finishing plaster

C after fixing plasterboard

D after mixing backing plaster

13 Which of the following backgrounds is three coat plasterwork applied to?

A plasterboard ceiling

B blockwork walls

C expanded metal lath

D concrete floors

14 What is the result of applying excessive thicknesses of undercoat plaster?

A drying

B sweating

C blistering

D sagging

15 Complete this sentence. Plaster ordered for a specific background must be:

A desirable

B suitable

C compatible

D economical.

16 What tool is used to remove rough edges of plasterboard after cutting?

A utility knife

B pad saw

C tape measure

D rasp

17 What tool is used to hold plasterboard in position when fixing to ceilings?

A gauge

B strut

C ledge

D whisk

18 What tool is used to cut gaps for services in plasterboard?

A panel saw

B circular saw

C hack saw

D pad saw

19 When fixing plasterboards by direct bond application, what are the vertical dab centres?

A 300 mm

B 400 mm

C 500 mm

D 600 mm

20 How many minutes of fire protection do fireline plasterboards provide?

A 30 minutes

B 60 minutes

C 90 minutes

D 120 minutes

CHAPTER 4

APPLYING EXTERNAL PLAIN RENDERING

INTRODUCTION

External rendering is another aspect of your plastering career that you will need to practise to develop. This type of work will test your skills, when you have to produce straight, flat and smooth surfaces that also contain sharp details to returns of angles and openings.

This chapter covers the materials, tools and equipment required to prepare and apply external rendering. It also explains the procedures and techniques you will need to develop to master this challenging skill.

By reading this chapter you will gain knowledge of:

1 what is meant by external rendering
2 how to interpret information for external work
3 how to select and prepare materials, tools and equipment for external work
4 how to apply render to external backgrounds.

The table below shows how the main headings in this chapter cover the learning outcomes for each qualification specification.

Chapter section	Level 1 Diploma in Plastering (6708-13)	Level 2 Diploma in Plastering (6708-23) Unit 224	Level 2 Technical Certificate in Plastering (7908-20) Unit 203
What is external rendering?	n/a	Learning outcomes 3 and 4	Topics 1.3, 2.1
Interpret information	n/a	Learning outcomes 1 and 2	Topics 1.1, 1.2
Select tools, materials, tools and equipment	n/a	Learning outcomes 3 and 4	Topics 2.2, 3.1
Apply render to external backgrounds	n/a	Learning outcomes 5 and 6	Topics 2.3, 3.2, 3.3

1 WHAT IS EXTERNAL RENDERING?

The purpose of external rendering in the building industry is to provide:
- a desirable finish that will enhance the appearance of a building
- a protective surface, preventing passage of moisture that can penetrate the external wall and enter the building
- a thermal barrier that enhances U-values of buildings and reduces energy consumption.

ACTIVITY

Research on the internet what is meant by U-values and how they are used to measure the energy performance of buildings.

There are many different types of external render finishes. The following table shows the different types of external surface and texture that are commonly used in the construction industry. Many of these render finishes are covered in greater detail in Level 3 Plastering qualifications.

External render finish	Description
Plain face 	Plain face finish is completed by scouring and **consolidating** the surface of render that has been ruled with a straight edge and flattened with a darby. It has a flat, smooth, sandy look.
Tyrolean 	A textured layered stipple type finish achieved over several passes to avoid **slumping** of material. The finish is honeycomb-like and applied to plain face backgrounds with a Tyrolean spray gun.
Ashlar 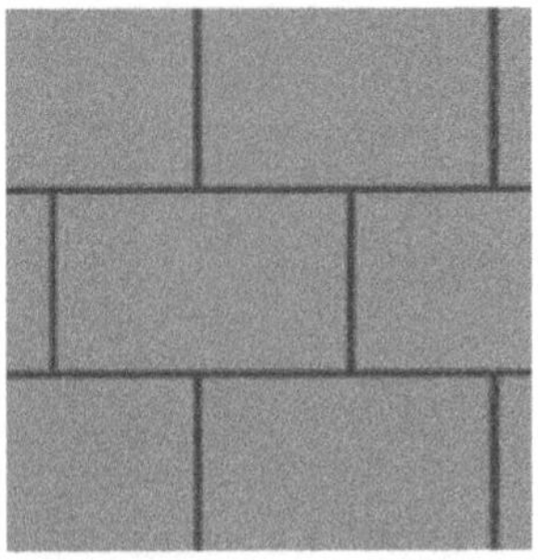	Ashlar is formed with a jointer or similar tool, carving out the shape of blockwork in the surface of plain floated render.
Brush 	Applied render finished by **rotating** a large bristled brush flat to the rendered face.
Pebble dash 	Granite, stone or flint spar thrown onto a buttery render mix which is applied to a **scratch coat** surface. (Buttery is the consistency of the mixed render for this type of application, to receive pebble dash finish; also known as dash receiver.)
Rough cast 	A mixture of granite and render material that can be thrown onto a **butter coat** or **keyed surface**.

External render finish	Description
Scrape texture	Achieved by scraping the surface of a pre-mixed render that contains additives or is polymer-modified.
Thin coat render	This external render is applied and finished by scouring the surface with a specialist-made float, forming a textured surface. It can be more efficiently applied using a spray hopper gun. Also known as 'light texture' and 'rubbed texture', it contains resin/acrylic/silicone as a binder, allowing the surface to be flexible and crack-resistant.

KEY TERMS

Consolidate: to close in the surface of a floating coat, render or floor screed with a float, making the surface flat, dense and compact.

Slumping: when plaster has been applied too thickly and slides down the wall due to excessive thickness and weight.

Rotating: small circular movement when applying brush textured finish.

Scratch coat: a plaster or render mix applied to a surface to control suction and provide adequate key before a floating coat is applied.

Butter coat: the top coat render mixed to a buttery consistency that is applied to receive dry or wet dash finish onto its surface.

Keyed surface: a surface able to receive an application of plaster/render that enables suitable adhesion of two surfaces.

HEALTH AND SAFETY

Many rendering products can be harmful. Always wear adequate PPE and read the manufacturers' technical information. Minimum PPE will be hard hat, safety glasses, dust mask, gloves, hi-viz jacket and safety boots.

Spray machines are often used today as a fast and efficient way of applying render. Ritmo L Plus machines are used mainly for projection plaster/render pre-blended bagged materials.

▲ Figure 4.1 Ritmo L Plus machine

Spray hoppers are used mainly for application of thin coat renders/Tyrolean Cullamix.

▲ Figure 4.2 Spray hopper

▲ Figure 4.3 A jointer is used for forming an ashlar pattern

▲ Figure 4.4 A darby is used for flattening and smoothing render

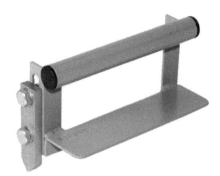

▲ Figure 4.5 The ashlar cutter tool by Refina. A jointer used for forming deep V and square ashlar joints

▲ Figure 4.6 A jointer used for forming a pattern with its teeth for ashlar joints

ACTIVITY

Search the internet for more types or makes of projection plaster machines that are widely used within the plastering industry.

Pulley wheels are sometimes used for lifting heavy buckets of mixed render onto scaffold lifts. On larger sites this would be done by the telescopic handler machine driver.

▲ Figure 4.7 Pulley wheel

▲ Figure 4.8 Telescopic handler

1 Research two types of render on the websites of manufacturers such as Sto, Weber and K Rend.

2 Look at types of background preparation for different substrates and identify the materials to use for background preparation.

3 Look at the types of topcoat finish available and match materials which are compatible (can be used together).

4 Research the correct beads and trims that are used in various render finishes, in particular the difference between insulated render finishes and other types of render system.

There are also other, less common, types of render that are used on older buildings. These are generally made to suit the character of the building and can create a very effective appearance; a good example of this is **cottage finish**. These types of render require less technique to apply and finish than other renders.

External renders are popular and can be used on a range of different backgrounds, including insulation. External wall insulation (EWI) is fixed to solid backgrounds mechanically and by direct bond, improving the insulation of the building.

▲ Figure 4.9 Insulation being fixed to a background, showing mechanical and direct bond fix together

Pre-blended bagged renders are preferred on insulated backgrounds because render contains polymers and is applied with reinforcing fibre mesh scrim cloth bedded into the render to reinforce the surface. Monocouche 15 mm application is not compatible with an external wall insulated system.

HEALTH AND SAFETY

Always wear protective gloves when handling fibre mesh scrim cloth to prevent cuts to hands.

This chapter focuses on plain face smooth render (see pages 181–187), which is best applied and finished on a scratch coat surface. This type of finish consists of several layers applied onto a solid background surface. It can involve either two or three coat plastering/rendering work, depending on the unevenness of the background.

INDUSTRY TIP

Some modern pre-blended bagged materials can be applied in the same day in two passes of 6 mm each, when applied to new concrete blocks with moderate suction.

If the background suction is too high, the render application will dry out too quickly. This will cause it to **eggshell** and shrink, causing **delamination**.

If the background suction is too low, the render application will not adhere to the surface, causing slumping and delamination.

Research the types and strengths of concrete block which are best suited to plain-faced render. You might find it helpful to visit the British Board of Agrément (**BBA**) website or read BS 5628 Part 3 Masonry.

KEY TERMS

Cottage finish: a traditional render finish applied in rustic fashion.

Eggshell: when plaster or render dries out too quickly, shrinks and cracks.

Delamination: when plaster or render becomes detached from a background and falls off, due to the eggshell effect.

BBA: within the construction industry, British Board of Agrément certification indicates a high quality, experienced and reliable company or product. It is highly regarded and used by manufacturers in industry as a symbol of quality.

Preparing and forming this type of finish will challenge your skills as an operative. Being eager, enthusiastic and keen to practise will help you to learn and develop the required techniques to apply and form the finished surface. You will need to produce high standards of workmanship, using a variety of hand tools to produce the final product.

To do this successfully, you will need to deal with factors such as **background key**, which helps to make the render adhere to the surface. You will also need to control and treat suction to apply the first scratch coat successfully. Another important factor will be the selection of suitable materials that will be compatible with the background, to ensure there is good adhesion and bond between the two surfaces.

The process of rendering includes laying on several applications of render. The number required will depend on the **specification**, which will state the background's properties. The specification might state that the background must be built up using dubbing out coats, before the scratch coat and top coat are applied. See the next section of this chapter for more information about specifications, **British Standards** and other sources of information.

▲ Figure 4.10 An uneven external background that requires dubbing out

KEY TERMS

Background key: the background surface; you may need to form (or 'key') a compatible surface to allow adhesion between various coats of plaster/render, either by using SBR/PVA application, or by forming a scratched, rough surface between application coats to enable the coats to adhere without delaminating.

Specification: instructions stating the standards required and practice to be followed for a task, usually BBA-approved and to meet British Standards. It is often an official document from the architect who is overseeing a project.

British Standards: standards produced by the British Standards Industry (BSI) Group, which is incorporated under a royal charter and formally designated as the National Standards Body (NSB) for the UK.

▲ Figure 4.11 Dubbing out stonework

2 INTERPRET INFORMATION

To carry out the work correctly, you will first need to prepare properly by interpreting drawings and reading various contract documents such as specifications, schedules and data sheets. You will need to extract from them the information required to carry out the different aspects of the work. More information on these documents is given in Chapter 1.

You can then use the gathered information to select appropriate tools and equipment, evaluate how to prepare the background surface and prepare for mixing and gauging the materials.

Scale: 1:200

▲ Figure 4.13 Block plan

> ### INDUSTRY TIP
>
> As a trainee, you will generally work from verbal instructions communicated by a person in higher authority – this could be the plasterer, supervisor or manager.

Working drawings

Learning how to accurately read drawings such as **block plans** will give you the necessary skills to develop and progress within your plastering career. Drawings provide accurate information about and descriptions of the rendering process and will describe the desired outcomes of the project. They illustrate building **elevations**, positions of pre-made profiles, sections of windows, door heads and reveals and also any special design features that might be included with the rendering work.

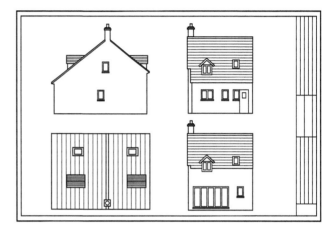

▲ Figure 4.14 Elevation drawings for a typical house

When you work on external rendering contracts, the architect's drawings provide detailed measurements of external wall surfaces. You can use these to calculate areas and volumes of material, as well as linear measurements for pre-made beads or trims. This will help you to schedule materials and labour costs if you are required to **tender** for the work.

Specifications

You can use the specification to help you order the correct materials and components for rendering work. The specification provides detailed descriptions of the materials that must be used for the work. It also contains other vital information, including the:

- type of background
- surface preparation
- mix **ratios** and additives
- required standards of workmanship
- method of application
- thickness
- tolerances
- types of pre-formed beads or trims to be used.

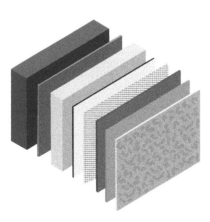

▲ Figure 4.12 Section of external wall and rendering systems

KEY TERMS

KEY TERMS

Block plan: drawing that shows the proposed development in relation to the surrounding properties.

Elevations: drawings that show the external walls of the building from different views.

Tender: to submit a cost or price for work in an attempt to win the contract.

Ratio: the proportion of materials mixed together; for example, 6 parts of sand to 1 part of cement would be written as 6 : 1.

Not following drawings and specifications might lead you to carry out the work incorrectly. There might be other trades working alongside you and any hold-ups in your work schedule will no doubt have a knock-on effect on their work too.

If you find any discrepancies (errors or things that do not match) in the drawings or specification, you must report them to someone in higher authority as soon as possible, to avoid disrupting and delaying the work. For more information on specifications, refer back to Chapter 1, page 31.

ACTIVITY

Write a specification for the image used in the 'Improve your maths' activity on page 164. Make sure the specification covers the points in the list above. Your tutor will help you with this task.

Schedules

The schedule is another document used in construction contracts, containing information on the amount of materials and components required for the work. For example, the schedule will list the amount of each type of pre-formed bead required for a rendering contract.

For more information on schedules, refer back to Chapter 1, page 32.

Manufacturer's information

Manufacturers' data sheets contain important information about the storage and use of their products, as well as product information and technical guidelines. For more information on data sheets, refer to Chapter 6, page 227.

You will need to comply with **manufacturer's instructions** and guidelines when using rendering products and components in order to maintain the quality of the completed work. Failure to do this can lead to costly mistakes, which can have a major effect on the outcome of the rendering work.

ACTIVITY

Search the internet to find a data sheet for cement and lime. List the information covered.

HEALTH AND SAFETY

The documentation is likely to include Risk Assessments and Method Statements (RAMS). RAMS are applied to most activities within the construction process and must be read and understood before any work is started, to enable safe working practice.

Calculate quantities of materials

At the start of every job, you should be aware of the possible effects of poor planning. Every effort should be made to ensure consistently high quality work and these efforts need to be maintained throughout the rendering process.

Always store materials correctly prior to use, so they do not become damaged, contaminated or damp/wet. If plaster/cement bagged products become wet or damaged before use, they will be unfit for purpose and must be disposed of in the correct recycling skips.

Check materials for quality and **shelf life** before using them: using poor quality materials can cause problems and will reflect badly on your workmanship and professionalism as a qualified operative. You may also incur the cost of having to redo or repair the work.

KEY TERMS

Manufacturer's instructions: these state what a product may be used for, how it is to be installed and the conditions it can safely be exposed to.

Shelf life: the use-by date of products such as cement and lime.

▲ Figure 4.16 Buckets are used to measure materials

▲ Figure 4.15 Poorly set-up mixing area and materials

It is good practice to order just enough materials to start and proceed with the job. Good planning and storage control, with order points set against your work schedule, will prevent you from running out of materials during the work. If this happens, it not only causes delays but can become costly if there are no materials for the hired labour to work with.

When pricing for work, you will have to calculate how much of each material you need for the area to be rendered. You can do this by measuring and calculating the wall area from the drawings. Once you have the measurements and calculations, the next stage is to find out the ratio of the mix. This will allow you to work out the amount of each material that you need. Reading the specification will also help you to do this.

Ratio of mixing materials

An example of a ratio of mixing materials is: 4 parts sand, 1 part cement, 1 part lime = 4:1:1.

Always use containers of the same size to measure the materials to be used, so the mix ratio keeps the same consistency all the way through the process. Always fill the buckets exactly to the top. Failure to do so will result in an inconsistent mix and a finished rendered substrate which has an appearance of differing shades.

EXAMPLE

- One 25 kg bag of sand will cover 4.5 m² at 10 mm thickness.
- One 25 kg bag of cement will also cover 4.5 m² at 10 mm thickness.

If the mix required is 4 sand to 1 cement (a 4:1 mix), we can calculate that 5 bags will cover 22.5 m² of wall at 10 mm thickness. The calculation for this is the total number of bags multiplied by the area covered by one bag:

$$5 \times 4.5 \text{ m}^2 = 22.5 \text{ m}^2$$

If you have a wall area that measures 225 m², how many 25 kg bags of sand and cement will you need to apply a 4:1 ratio of sand to cement at 10 mm thickness? To find the answer, divide your wall area by the area covered by the 4:1 mix:

$$225 \div 22.5 = 10$$

So you will need 4 × 10 = 40 bags of sand and 1 × 10 = 10 bags of cement to complete the job.

To allow for minor errors, always add around 10% extra for wastage. You need to work this out for both the sand and cement.

- Sand: 10% of 40 is 4.
- Cement: 10% of 10 is 1.

So you need another 4 bags of sand and another 1 bag of cement. This brings the total for the job to **44** bags of sand and **11** bags of cement.

Formulas

You need to know a number of formulas to calculate volume. For example, when rendering around the top of a house, you will need to know:

1 the formula for calculating the volume of the triangular part of a **gable apex**:

half-length × height × thickness

2 the formula for calculating the volume of the square or rectangular part of the gable wall:

length × height × thickness.

If there are any window openings, you will need to subtract the volume of the openings from your gable wall volume.

You can calculate the quantity of beading required from the **linear measurements** of openings.

KEY TERMS

Gable apex: the triangular part of a gable wall.

Linear measurement: measurement of a straight distance between two points.

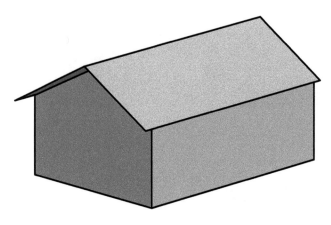

▲ Figure 4.17 A building with a gable-ended roof

IMPROVE YOUR MATHS

Answer the following questions, using this drawing.

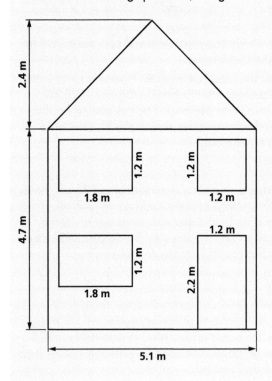

How many 25 kg sand and cement bags will you need to cover the gable elevation, with a ratio of sand cement render at 4 sand to 1 cement at 20 mm thick? Round up your answer to the nearest number of bags.

Remember from page 157 that:

- one 25 kg bag of cement will cover 4.5 m² at 10 mm thickness
- one 25 kg bag of sand will also cover 4.5 m² at 10 mm thickness.

To find the answer you will need to do the following:

1 Work out the area that one bag of sand and cement covers at 20 mm thick.

2 Work out the area to be covered. This will involve working out the gable area (the triangular part) and the rectangular part of the wall and then adding the two together.

3 You will then need to subtract the area of the four openings from the area to be covered.

4 Divide the area to be covered by the area that the 4 : 1 mix covers.

5 Add 10% extra for wastage for both sand and cement, rounded up to the nearest whole bag.

③ SELECT MATERIALS, TOOLS AND EQUIPMENT

Rendering materials

Selecting and preparing the correct materials plays a major part in the external rendering process. Knowing what materials to select and how to prepare them will allow you to plan your work methodically.

You need to refer to the specification when selecting materials, as this will provide detailed descriptions of the required types.

Cement

Cement is used as a **binder** and to provide strength in the mix. It is made from 75% limestone and 25% clay and is used because of its faster setting time compared with lime mixes. Cement mixed with sand and water will begin to set after 45 minutes and will normally be completely set by the next day. Cement-based mixes can take several days to reach their final strength and this process is known as **curing**. Ordinary Portland Cement is one of the most commonly used cements in mortar.

White cement is another type of cement, which is used in the render finish coat to enhance its appearance, providing a much lighter look. It is made from limestone (75%) and white china clay (25%). Gauging white cement for rendering should be done accurately to ensure the mix is uniform in colour and consistent in strength.

> **ACTIVITY**
>
> Research how white cement is manufactured and discuss your findings with your tutor.

▲ Figure 4.18 Bagged cement

> **KEY TERMS**
>
> **Binder:** a material used to make the aggregate stick together when mixed.
>
> **Curing:** allowing the mix to set and reach its full strength.

Lime

Lime is another material that has many uses in plastering mixes. It is made by crushing limestone and heating it in a kiln. Lime comes in bags and should be stored on pallets in dry conditions. There are two types used in plastering mixes: hydrated lime and hydraulic lime.

- Hydrated lime does not set when added to plastering mixes. This has many benefits. Lime added to a cement-based mix will improve the mix's workability and adhesion and help prevent shrinkage. Another benefit of lime is that it creates suction.
- Hydraulic lime has a 'chemical set' and sets very slowly when mixed with sand. The setting times can be influenced by the temperature – setting will take longer in winter than in summer, so good planning of the work is required.

▲ Figure 4.19 Bagged lime

Sand

Sand is an aggregate used to bulk the mix. Without sand, cement mixed with water would be too hard and would crack.

Many different types of sand are used in construction. Sand used for rendering should contain a mixture of small, medium and large grains and contain as little silt as possible. Silt is a very fine grain that can prevent the cement from binding the mix, resulting in the render surface becoming weak and powdery.

Some types of sand should be avoided because of the shape of their grains – they can be too round in shape and the sand will therefore contain too many **voids**.

- Pit sand is generally used by bricklayers for building work. This sand is very fine with round-shaped aggregate and contains too much silt for rendering.
- Silica sand and silver sand are deemed too fine for plain face rendering. Instead, they tend to be used as a fill-in between paving bricks and for filling in fine seams or joints between the masonry.
- Loam sand is a term used for sand containing clay. This type of sand is used for traditional lime-based renders in the restoration of old buildings.

▲ Figure 4.20 Sand

Additives

Additives play an important part in the render mix and you need to know when and how to use them. They usually come in liquid form in 5- or 25-litre containers. The table below lists the different types of additives and how they can enhance the performance of the render mix.

INDUSTRY TIP

Sea sand contains salt and impurities such as shells. It is considered badly graded so should never be used for rendering.

KEY TERMS

Voids: pockets of air, common in poorly graded sand.

Additive: a substance added to plaster mixes to change their natural properties.

Additive	Use
Plasticiser	Plasticiser creates air bubbles in the mix and is used to improve the workability of the mix. Without plasticiser, the mixed material would separate and become too heavy to spread. Too much plasticiser in the mix, however, can make it weak and cause the mix to crumble.
Waterproofer	Added to the mix to prevent water penetrating through.

Additive	Use
Frostproofer/accelerator 	Added when cold or frost conditions are expected during the work, to speed up the setting time.
Salt inhibitors 	Used to prevent the effects of **efflorescence** from penetrating through the background surface.
SBR (styrene-butadiene rubber) bonding slurry 	Used to increase the bond on poorly keyed surfaces.
Mould remover 	Mould remover is not an additive but is used on backgrounds or render surfaces that have been affected by mould growth such as fungi, mildew, moss and algae. Mould remover is painted or sprayed onto the affected surface to remove and destroy the mould spores and other airborne particles.

KEY TERM

Efflorescence: a white powdery deposit on the surface of plaster, containing a high proportion of salt.

ACTIVITY

The plasterer's labourer has added far too much plasticiser to the render mix. Research what would happen to the mix and report on what steps must be taken to prevent this from happening again.

Manufacturers' guidance and instructions should be followed when using additives. Measuring the exact amounts to add to clean water is vital: adding too much or too little will cause the mix to lose its strength and adhesion properties.

ACTIVITY

In groups, look at the different types of additive used for external rendering. Read the instructions on their use and feed back your findings to the whole group.

Storing materials

You will need to prepare designated areas for storing bagged materials and containers. Poor housekeeping and stock control can result in waste. Consider a solid flat surface for loose materials such as sand and make sure it is protected with a good heavy cover. Using **contaminated** materials can negatively affect the work and it will be costly to replace or renew them.

New deliveries of materials must be checked to make sure the delivered goods match the order in terms of quantity and quality. Check use-by dates before signing the delivery note. If any materials are not accounted for, or are damaged or out of date, record this on the delivery note and report it to your supervisor as soon as possible. You can check the shelf life of bagged material by looking at the use-by date on the bag.

> **INDUSTRY TIP**
>
> Always put new deliveries at the back of the stock pile, to ensure the older materials are used first. This is known as **stock rotation**.

Storing sand

Sand can be purchased loose, in bags or in sacks. Sand used in plastering mixes should be protected from leaves or animal contamination, as this can affect the binding of the material and the quality of the surface finish. A good cover or tarpaulin is useful to protect from rainwater (which can cause **bulking**). Bulking increases the weight and volume of sand, which leads to inconsistency when gauging and mixing the render mix.

▲ Figure 4.21 Sand in bays

> **KEY TERMS**
>
> **Contamination:** when materials have been in contact with something unclean, such as leaves blown into the sand or dirty water used for mixing.
>
> **Stock rotation:** ensuring old stock is used before new stock. When new stock is delivered, it should be stored behind the older stock, which needs to be used first.
>
> **Bulking:** the swelling of sand when it is wet, making it heavier.

> **INDUSTRY TIP**
>
> Some plasterers doing three coat work will add coarse sand to the dubbing out coat to give it strength.

Storing cement

Store cement off the ground, under cover and away from damp conditions. Contaminated or out-of-date cement will become hard and lumpy and result in a weak set. If accidentally used in the render mix, it might not bind the sand and the render will lose its initial strength. Using poor quality cement might also cause the surface of the render to become powdery and soft.

If cement is stored for a long time, even in good conditions, too many bags on top of each other can cause the cement to set in the bag, becoming unusable. This is known as pack set.

▲ Figure 4.22 Bagged cement in storage

Poor quality materials can result in loss of strength, causing mixes to break down and lose their ability to bond and be compatible with the background.

Make a list of the effects of poor storage of materials such as cement, lime and sand and how they can affect the render mix or surface.

Selecting pre-formed beads/trims

You will need to select the appropriate type of pre-formed beads for your rendering work. Each one has a specific purpose and use and should be fixed in the correct position.

Type of bead	Use
Angle beads 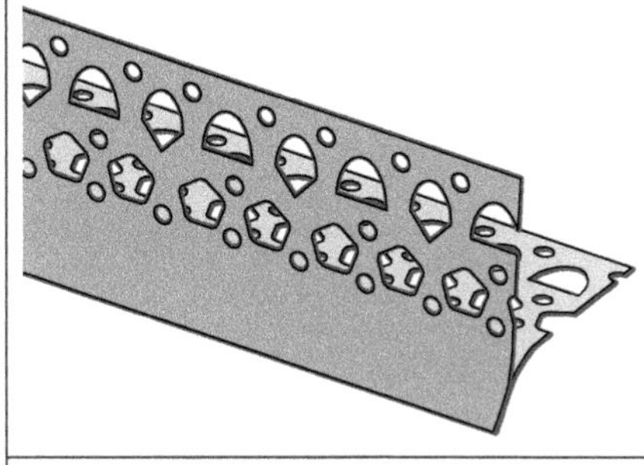	Used on corners. They form straight **arris edges** and reinforce vulnerable corners by protecting them from impact.
Stop beads 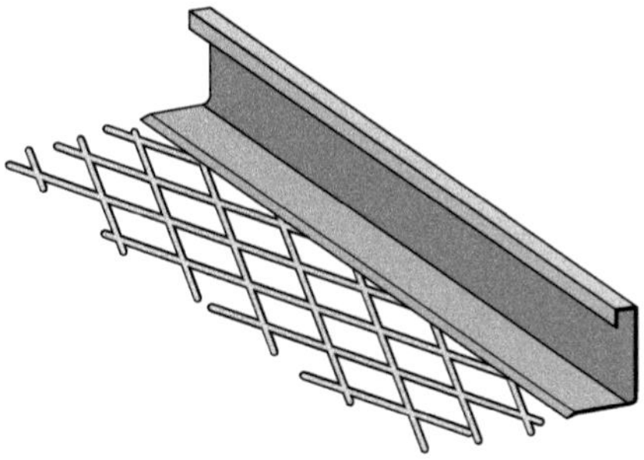	Used when a straight stop edge is required. This could be for rendering up against different surfaces, such as cladding or facing brick.
Expansion beads 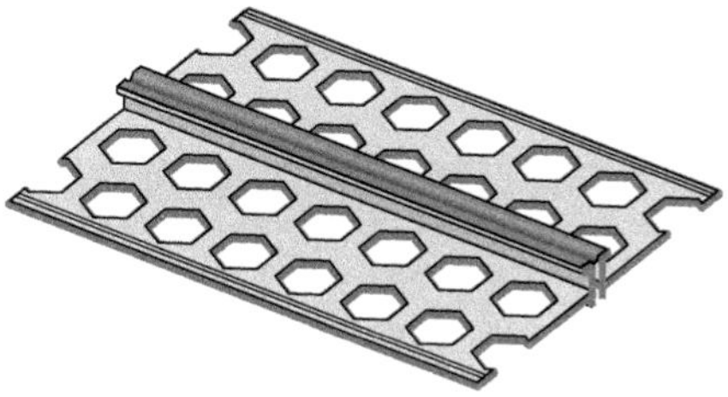	Used along straight joints in brickwork or blockwork, allowing for slight movement and preventing cracking.
Render/bell beads 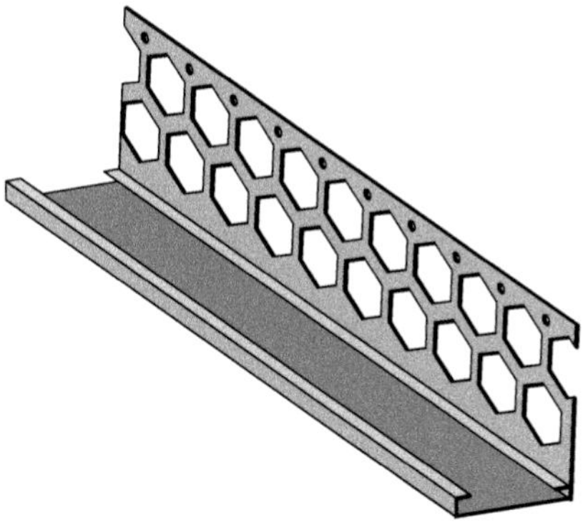	These are fixed to bridge the **damp proof course (DPC)**, window and door heads to act as a drip. They are also used to break down large areas of render into more manageable working areas.

Type of bead	Use
Reveal bead with mesh	Used to allow expansion and movement against a window frame to reduce cracking direct to a window.

KEY TERMS

Arris edge: a corner feature formed as a sharp edge finish with angle bead trim or by forming a hard angle.

Damp proof course (DPC): a layer or strip of watertight material placed in a joint of a wall to prevent the passage of water. Fixed at a minimum of 150 mm above finished ground level.

Beads are manufactured either in stainless steel or plastic and are available in various lengths and thicknesses. The different sizes allow for their use when applying render of different thicknesses. Fixing and positioning beads is explained later on in this chapter.

ACTIVITY

Carry out research into different manufacturers that produce beads and trims, for example: Expamet, Catnic, ProBead and Renderplas.

Selecting tools and equipment

Basic tools and equipment, such as the trowel and hawk for laying render on the background surface, will be familiar to you by now. You can refer to Chapter 3 for more on the use of these tools. However, other tools and equipment are also used to carry out rendering work.

Tool	Use
Straight edge/feather edge 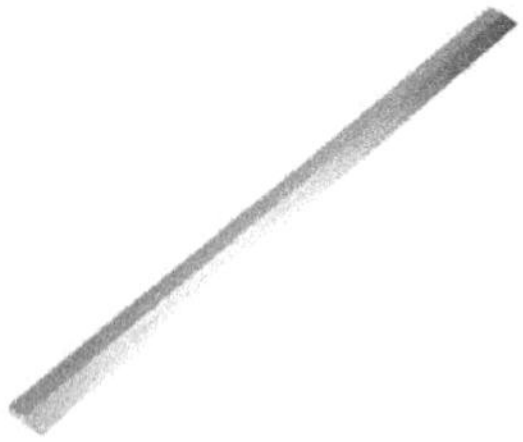	Straight edges are used to check backgrounds and rule the render surface, removing any high points and filling out low points. This tool can also be used to form the external angles of the render surface and the corners of returns.
Darby 	Used for flattening and ruling surfaces. You will need to be competent to use this successfully. It can also be used to form the corners of returns.

Tool	Use
Comb scratcher 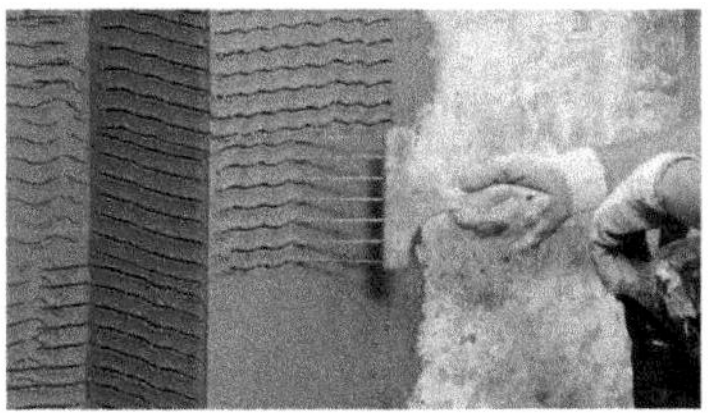	Used to key the surface of backing coats.
Float 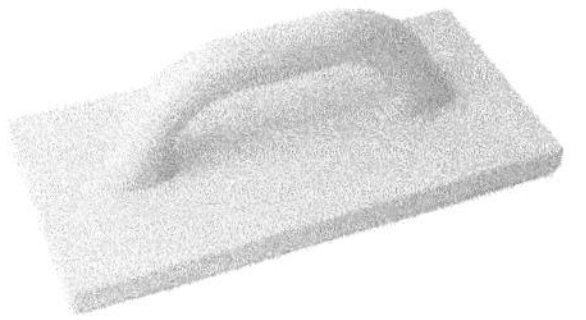	Used to consolidate the final surface of the render finish, removing any high points and filling in low areas in the render surface. A wooden float was traditionally used to carry out this work; plastic floats are now often used.
Tin snips 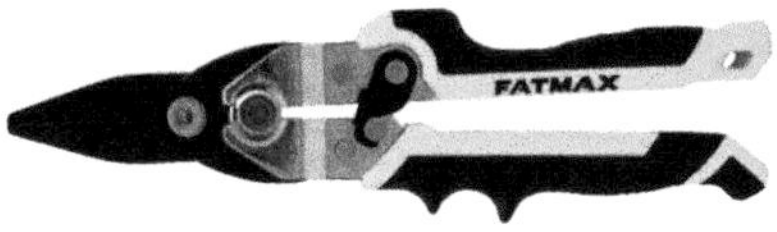	Used to cut beads to the required length.
Drum mixer 	A strong, mechanical, robust mixer is required to make the render.
Serrated straight edge 	Used to rule the render surface and reduce air pockets in the render.
Finishing spatula 	Used to consolidate surface after use of serrated edge.
Sponge float 	Used to consolidate the final surface of the render after use of the float.

➡

Access equipment

When rendering, it is important that the scaffold platform is suitable for the job in hand. The table below details some of the equipment you may use.

Equipment	Use
Independent scaffold 	A good type of scaffold to use when rendering outside surfaces because it is erected away from the wall. It is a wide and solid working platform, allowing plenty of room to carry out your work.
Trestle staging 	A good scaffold for low level buildings, which can be erected and dismantled with ease. A firm, flat base is always required for erecting trestles. Some sites will not allow trestles to be used. You should always check what type of access equipment can be used on the site where you will be working.
Hop-up 	A good staging to reach low levels up to roughly 2.4 m in height.
Podium access equipment	Often used on larger sites when trestle staging and hop-ups are not allowed.

Equipment	Use
Mobile tower  	Often used on smaller elevations.
Semi-electric scissor lift 	Often referred to as a mobile elevating working platform (MEWP), this is used for quick and efficient access on lower elevations.

ACTIVITY

1 Research three types of access equipment shown in the table.
2 Highlight any specialist training that might be required to safely use the equipment under the current Working at Height Regulations.

HEALTH AND SAFETY

SCAFFTAGs are used on equipment from which a person might fall 2 metres or more. A scaffold should be checked before it is used for the first time and then every seven days until it is removed. Information on scafftags will show when a scaffold was last inspected and indicates whether it is fit for use.

▲ Figure 4.23 Example of a SCAFFTAG

ACTIVITY

1 Research different types of scaffold inspection system and report on the use of scafftags when working at height.

2 Look at various types of scaffolding for different types of contract, to see what is necessary by law when working at height. Websites that might offer advice include Smart Scaffolder and the Health and Safety Executive.

ACTIVITY

Make a list of tools and equipment you would need to apply two coat work to brick or block backgrounds that contain a **return**, require a **bell cast** above a window and require a **plinth** along the base of the wall at DPC height.

4 APPLY RENDER TO EXTERNAL BACKGROUNDS

Planning the work schedule

Before you start work on site, you need to read the RAMS to minimise risks involved with the work. This will help to prevent accidents and injuries that could occur during the rendering process.

Although method statements are produced to plan a safe method of work, there are other factors to consider before applying external plain face rendering to background surfaces:

1 Working at height causes many safety issues.

2 The climate can affect the process and cause damage to the surface of the work:
   - Sometimes weather conditions change in minutes and can have an instant effect, causing a wash-down of the applied surface from rain, or causing surfaces to dry too quickly when render is applied in direct sunlight or windy conditions.
   - In winter you need to be aware of frost and freezing conditions, as these will cause the mix to become weak or crumble as the material thaws out.

3 On some renovation work, you will need to consider and plan how to deal with television aerials, satellite dishes, telephone cables, services and pipes. These should be removed while the rendering work takes place and then reattached after the work is complete.

4 Drains and gullies should be covered and protected to prevent blockages. Windows, doors and ironmongery should be covered with sheeting or cling film to protect them from the render mix.

5 Unauthorised entry and pedestrian traffic areas are important factors to consider when planning your work. Display signs and barriers to warn other operatives of possible dangers and that the rendering work is in progress. This will also help to prevent accidental impact damage to the surface of the work, which would be difficult to repair and costly to replace. Use **hoardings** or fences to cordon off the work area, preventing any trespassers or intruders from entering the work area.

INDUSTRY TIP

Always check the latest weather predictions before you carry out rendering work, in case the weather is likely to affect the finished work.

Preparing background surfaces

The background surface determines many aspects of the rendering process. It forms the base for the render and if it is deemed weak with a poor key it will not be compatible with the applied mix. It is important to understand that not all backgrounds have the same properties and different methods and techniques are required to prepare them.

Backgrounds can:
- be soft and weak
- be hard and dense
- be uneven, requiring building out
- have low suction or high absorption rates.

Checking the suction and absorption will tell you if the background is dry and porous. You can find out by applying water to the background with a splash brush, to see how quickly or slowly it is absorbed. No or low suction will indicate that the background is hard or dense.

Two coat backgrounds are classed as flat and straight and are finished using two applications of render. Some uneven backgrounds require an additional coat of render; this is termed **three coat work** (dubbing out, scratch and finish).

▲ Figure 4.24 Applying water with a flat brush

Check the background's surface before starting work and remove any mortar snots that might have been left by the bricklayer when they carried out their brick work or blockwork.

KEY TERMS

Hoarding: a barrier surrounding a site to protect against theft and unauthorised entry.

Three coat work: when plastering exteriors, this means applying three distinct layers of render: dubbing out/pricking up, scratch and finish render surface.

Background types

Let's examine the surface characteristics of different backgrounds in more detail.

Type of background	Comment
Composite backgrounds	Composite backgrounds (backgrounds made up of two or more materials) should be prepared with stainless steel **expanded metal lath (EML)**, fixed using **mechanical fixings** and plugs. This will reinforce the render when it is applied to the background, creating a strong surface.
Stone or slate backgrounds	This type of background cannot be completed in two coats due to its unevenness, with deep crevasses and recesses that will require dubbing out. Stone and slate have poor key and should be prepared by application of a bonding slurry on their faces to improve adhesion when rendering.
Clay bricks	Clay bricks were very popular at one time and can be found in all types of building. A common fault with clay bricks is that they shell their face, causing the plaster/render to 'blow' (come away from the background). This background is often uneven because the bricks were manufactured in kilns at great heat which made them all a slightly different shape. They were then laid on a lime mortar bed, which is very weak. Clay bricks and lime mortar joints have a high absorption rate that will cause high suction levels. This surface should be treated with a bonding adhesive before plastering/rendering. Raking out the joints will also improve the key.
Concrete bricks	These bricks are made from concrete aggregate mixed with cement. This surface is smooth and hard, which means that the key is poor and the suction is minimal. A bonding agent is best suited for this surface.
Engineering bricks	This is a hard, dense surface with poor key and no absorption rate. The face of the brick has a glossy surface that makes it difficult to prepare for plastering/rendering. It has an enamel look and no suction. This surface should be scabbled or roughened to remove the sheen and then an **external slurry** can be applied. Alternatively, you can fix sheets of EML to its surface with mechanical fixings – this is a good way to reinforce and form a key on the background.

Type of background	Comment
New blockwork 	Newly constructed buildings that have block walling need little preparation before you apply plaster/render to their surface because they have medium to adequate key. Water can be applied in warm humid conditions to prevent the render from drying out too quickly. The surface is flat and can be rendered using traditional or modern pre-mixed materials. Block walling built to today's specifications and standards needs only a scratch coat and finish; this is known as two coat work.

KEY TERMS

Expanded metal lath (EML): sheet material in the form of diamond-shaped mesh that is used to reinforce a surface. This material can be fixed with screws and plugs or galvanised nails, or it can be bedded into the render material.

Mechanical fixings: fixings used to fix EML to composite backgrounds.

External slurry: thin, sloppy mixture of cement and bonding adhesive applied to a background to bond render to the surface.

Using bonding agents

There will be occasions when you need to prepare the background surface to improve its adhesive properties, otherwise the external render will not bond and will become loose. This could lead the render to develop severe cracks that become widespread over the surface.

The specification will provide specific details about the background surface and how it needs to be prepared before the render is applied. Instructions on mixing and applying bonding agents can be obtained from data sheets; you will need to follow these carefully to avoid breakdown between surfaces and so you do not invalidate the guarantee of the rendering work.

Some bonding agents can be applied to surfaces and left with a textured surface, to form a good key.

ACTIVITY

List five areas or surfaces that would need to be protected on an external contract before you apply slurry.

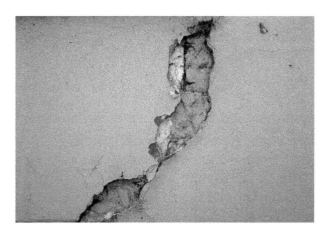

▲ Figure 4.25 External render with surface cracks

SBR bonding slurry

SBR is a strong bonding agent that is mixed with 1 part SBR to 2 parts cement to make a bonding slurry. This type of slurry is applied to the background and while it is still tacky the scratch coat mix is applied over the top of it to ensure a good bond between the background and the applied material.

▲ Figure 4.26 SBR container

INDUSTRY TIP

Remember to protect surfaces before you apply the slurry. Splashes can be difficult to remove from glass and can also stain surfaces such as masonry, timber and plastic.

▲ Figure 4.27 Applying SBR slurry

▲ Figure 4.28 Applying spatterdash to a common brick wall to create a textured surface

Polymer-based bonding adhesives

There are many types of **polymer**-based bonding adhesive on the market today, specially designed to overcome problems with adhesion. They contain polymers that have been tested in laboratories to enhance their performance, providing added insurance and a manufacturer's guarantee.

Manufacturers' instructions and data sheets provide simple user guides with instructions on mixing and application. Failing to follow these will invalidate the guarantee and cause a breakdown in adhesion between the applied material and the background.

KEY TERM

Polymer: strong glue-like substance used to improve the adhesion of render surfaces.

A traditional method for bonding surfaces was to use a spatterdash slurry (see Figure 4.27). This is a mixture of cement and sharp sand, made into a slurry and then thrown onto the background with a paddle or small shovel. In more recent years, glues have been added to increase and improve the bonding properties.

The first coat of render applied to EML which has been mechanically fixed to a substrate is known as a pricking up coat.

ACTIVITY

Carry out a simple bonding test on a brick surface using different bonding agents.

1 Mix the slurry and apply it to the brick face, then apply the render while the slurry is still tacky and leave for three days to cure.
2 Repeat the process, but this time use the brush to form a stipple pattern with the slurry and leave it to dry on the brick face overnight.

Your tutor will help you set up this task.

Setting up equipment

Reading the specification will help you to set up before mixing and having the necessary tools and equipment to hand will make the process simpler to complete. You will need equipment for:

- mixing, handling and lifting, such as buckets and tubs
- transporting, such as wheelbarrows and shovels
- access, including scaffolding and ladders
- storing waste, such as skips or bins.

INDUSTRY TIP

Before you start the mixing process, make sure you have the necessary health and safety equipment and clothing (PPE).

ACTIVITY

Make a list of:
- dangers associated with mixing
- safety clothing that you will need to carry out this work.

▲ Figure 4.29 Equipment for external rendering: wheelbarrow, shovel and skip

The next stage is to prepare for mixing, which will include mixing and gauging different materials and additives to specified ratios.

Mixing equipment

A mechanical drum mixer (see page 171) powered by electricity or fuel is best for mixing render materials. Set up the mixer in a designated mixing area, away from traffic and pedestrian routes. A suitable place for this work is outside the building and as close to the work as possible, with good access for transporting the render mix.

The materials should be gauged and measured in buckets or a purpose-made box to ensure consistent quality and strength throughout the mixing process. Mixing materials such as cement and lime outside can help reduce dust inhalation and adheres to Local Exhaust Ventilation (LEV) health and safety good practice. You also need to wear the appropriate safety equipment to protect your lungs and eyes.

Water and power are essential for mixing efficiently. Storing materials such as sand close to the mixer will reduce physical labour and increase efficiency. Before you start the mixing process, make sure the mixer you will be using is well maintained and fit for purpose.

Mixing method

Render for two coat work would normally be mixed to the following ratios:
- Scratch coat: 4 parts of sand to 1 part of cement with waterproofer additive.
- Top coat: 5 parts of sand, 1 part of cement and ½ part of hydrated lime with a plasticiser additive.

INDUSTRY TIP

When applying several or subsequent layers of render, the first mix should be strong, followed by the same strength or a slightly weaker mix to avoid creating stress. Stress can lead to cracking and cause the render to blow from the surface.

The architect is responsible for designing and writing the specification and will decide on the ratio for the mix.

The following step-by-step instructions show the method for mixing by machine, using the ratio 4 parts sand to 1 part cement with waterproofer.

STEP 1 Set up the mechanical mixer in the designated area. Set up the other equipment and materials near to the mixing area.

STEP 2 Fill a bucket with clean water, add the required amount of waterproofer and mix.

STEP 3 Pour some of the water into the mixer.

STEP 4 Fill one bucket full of cement and add it to the mixer.

STEP 5 Fill four buckets full of sand and add them to the mixer.

STEP 6 Let the mix turn slightly dry rather than wet in consistency for a couple of minutes, so the mix becomes workable.

STEP 7 When the mix is ready, place it in a clean wheelbarrow and transport it to the work area.

STEP 8 Set up the spot board and stand and soak the board. Empty the mix from the wheelbarrow onto the spot board and you are ready to go.

Accurate gauging of materials and additives reduces the possibility of incorrect ratios, which cause poor surface finishes and variable strength and colour in mixes.

Pre-blended modern renders

Modern plastering methods and materials have changed the way in which plasterers work. Pre-blended/mixed renders in sealed bags are available in a range of colours that have been specially designed. These renders cause fewer problems than render mixed on site, because they have been manufactured and batched in specialised processing plants.

Common pre-blended renders are scraped texture finish and one coat render (OCR) finish.

- Scraped texture finish is hand applied to 18 mm thickness in two passes and scraped back around 16 hours after application, depending on weather conditions.
- OCR is applied in two passes of 8 mm to make 16 mm thickness and finished flat smooth float/sponge float finish on the same day, depending on weather conditions.

Both applications can be machine-applied using different types of projection plaster/render machines. Although expensive, these machines increase the application method hugely by applying more volume per day than applying by hand. They also save the wear and tear on plasterers' tools over a considerable amount of time.

Before being bagged, these products undergo vigorous checks to make sure they meet industry standards. Another benefit of using these renders is that they carry a manufacturer's guarantee (as long as they have been applied in accordance with the installation guidelines provided by the manufacturing company).

▲ Figure 4.30 Pre-blended render for external rendering

Applying two coat rendering

Let's look at the process of applying plain face rendering using two coat work on brick or block backgrounds. This is completed in three stages.

Stage 1: Apply the scratch coat

The first application of render for two coat work is known as the scratch coat. This is normally gauged and mixed using 4 parts sand to 1 part cement. A waterproofer additive is measured and mixed with the water to form a waterproof barrier in the render when it has set.

The scratch coat provides the base for the top coat and is applied about 9–12 mm thick. However, the thickness depends on the unevenness of the background. In some instances the mix can be made stronger, especially if exposed to open areas with severe climate conditions.

If you **rule** the scratch coat, this will make the base even and it will be easier to apply the top coat to an average thickness. It also helps the top coat to dry evenly, allowing you adequate time to form the finish. Be careful when applying the render because applying it too thickly can cause it to sag and slide.

KEY TERM

Rule: flatten off plaster/render using an aluminium darby/straight edge rule.

STEP 1 Check for suction by wetting the background. Set up the spot board and load it with material.

STEP 2 Using a low level working platform, apply the render material, starting from the right-hand side and laying on a trowel length at a time. Spread the render from side to side, flattening the surface to a thickness of roughly 10 mm.

STEP 3 Check the surface with a straight edge and remove any high points.

STEP 4 Once you have finished a section of the wall, key the surface using a comb scratcher.

STEP 5 After completing the wall, use a gauging trowel to remove any render that is left on the floor.

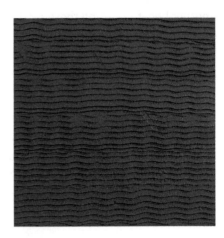

STEP 6 The completed scratch coat surface.

▲ Figure 4.32 Pre-formed beads

▲ Figure 4.31 Example of alkali resistant fibreglass mesh cloth

Once you have applied the scratch coat, clean the work area and leave the scratch coat to dry. You will need to let the render surface dry for a couple of days before you start the next stage.

Stage 2: Fix the bell bead

The next stage is to fix the different types of pre-formed beads or trims. Their positions will be shown on the working drawing. The total number of beads required for a contract will be listed in the schedule. Refer to the relevant data sheets as they will provide information on precise installation and fixing procedures.

The specification will give specified makes and designs of the beads. They should be made from plastic or stainless steel; galvanised products should be avoided for external work as the coating can corrode, causing rust stains in the work. In modern practice, most external render projects use plastic PVC external render beads.

Bell beads are used to form a bell cast. The purpose of the bell cast is to form a drip along the bottom of the wall, directing rain water away from the render surface below the DPC. On traditional renders, this practice is also carried out over window and door openings to direct water away from *in situ* frames. However, there is no requirement to do this with modern render systems as they have improved non-porous qualities and water will run off without penetrating existing frames.

There are two methods for forming the bell cast:
1 timber battens alone
2 timber battens with pre-made bell beads resting on top.

With both methods, first the timber batten has to be fixed along a level line. Then render is applied to the timber along the length, forming a slight curve on the wall about 150 mm from the approximate depth of the curve formed by the bell cast shape. This render is left to set overnight before the timber is removed.

If bell beads are used, a plastic or stainless steel bead is rested and fixed along the timber batten and then rendered to form the bell cast. Before fixing, tin snips and a tape measure are used to cut the bead to the required length and angle of cut.

Bell beads are also used above window and door openings for the same reason and provide a rigid fixed profile line to finish the render.

▲ Figure 4.33 Timber batten used to form a bell cast

▲ Figure 4.35 Completed bell cast in place

▲ Figure 4.34 Forming the angle at the bell cast with a timber batten

▲ Figure 4.36 Fixing beads along the timber batten

INDUSTRY TIP

Using a timber batten is a good way of making sure you have a straight surface to rest the bead on before fixing. Trying to fix the bead without a batten may cause it to buckle and distort along its length, which is unsightly to the eye.

STEP 1 Using a pencil, mark the position for fixing the bell bead on one side of the wall. This could be 300 mm off the floor.

STEP 2 Using a level and straight edge, transfer a level line from the marked point.

STEP 3 Fix a timber batten along the line.

STEP 4 Rest the bell bead along the timber. Then fix the bead using nails or screws or by bedding it into render mix.

STEP 5 Apply render to the bell bead and form the bell cast shape, cutting back about 10 mm along the edge of the bead to allow for the application of the top coat.

STEP 6 The completed bell cast, keyed with the comb scratcher.

Fir tree fixings are a more modern type of mechanical fixing for beads. They are pre-drilled to a depth and gently tapped in to fix with a lath hammer. It is good practice to bed pre-mixed external bead adhesive on top of fir tree fixings. Do not use pink grip adhesive to fix beads externally, as the pink grip might bleed/grin through the finished render.

INDUSTRY TIP

Do not mix and match different manufacturers' materials. If there is any kind of failure, all guarantees will be void due to this type of practice.

▲ Figure 4.37 Fir tree fixing

Forming timber bell casts above windows

There are several methods used to form drips above openings, such as windows and doors, using timber rules or roofing battens instead of pre-made beads. One method is to cut the timber and notch both ends, allowing the bell cast to project about 75 mm past the reveal.

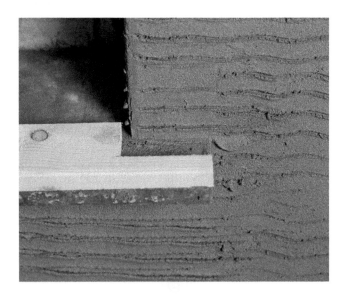

▲ Figure 4.38 Notching the timber

▲ Figure 4.39 Notched timber bell cast in position

The other method is to position a timber batten on its edge along the head of the window or opening.

▲ Figure 4.40 Timber batten fixed on edge

▲ Figure 4.41 Forming the bell cast up to the timber batten

With both of these methods, the bell cast must be supported from below by timber struts as it is formed. The **soffit** would then be finished plain and smooth once the timber has been removed.

Stage 3: Form the plain face finish

The top coat is the final coat and will need to be applied and ruled to **tolerances** and standards set out in the specification, which will test your skill. This type of finish will look poor if not ruled accurately; an uneven surface will cast shadows when the sun glares on its face. The same procedure for setting up should be followed as for Stage 1.

KEY TERMS

Soffit: the underside of a window or door opening.
Tolerance: required standards and accuracy of completed work.

▲ Figure 4.42 Top coat

The mix for this application includes cement, lime and sand. The ratio is normally 6 parts sand, 1 part cement and 1 part lime, with a plasticiser additive included to improve the mix's workability. Hydrated lime does not set but is added to reduce shrinkage and improve the mix's workability, making it fatty and good for spreading and ruling. Plasticiser also helps improve the workability of the mix; without this, the mix would be heavy, dense and very difficult to apply and spread.

You must apply your render to the correct thickness:
- If it is applied too thin, it will dry too quickly.
- If it is applied too thick, it will sag and be unable to line in when you rule it with your straight edge.

The surface of the render has to be rubbed and consolidated with a plastic or wooden float while the surface is starting to set. A good way to do this is to have some render on your hawk which can be used when you rub the face with the float, filling in minor defects or holes as you finish the surface. If you time this right, the surface will look straight, flat and plain.

INDUSTRY TIP

Take care when using additives with the mix: adding too much will weaken the mix and cause the face of the finish to crack or crumble.

If the render dries and sets inconsistently, it might be difficult to form the finish and achieve a uniform appearance to the surface. Poorly prepared surfaces that dry too quickly result in dark sandy patches that stand out on the face of the render, due to over-consolidating.

STEP 1 Apply the render mix onto the surface, working from the top right corner. Complete a section of work roughly 1 m² or an area that you can rule with a straight edge or darby.

STEP 2 Rule the surface with a straight edge or darby, filling in any hollows as the work proceeds.

STEP 3 Continue to work along the top of the wall, applying and ruling from your previously laid section.

STEP 4 Before you finish, check that the surface is flat and that it lines in with the straight edge.

STEP 5 Complete the bottom part of the wall just above the bell cast and proceed to the other end, using the straight edge or darby to rule and check that the surface is straight.

STEP 6 Apply the render to the bell cast, forming a slight angle in the wall's surface at this level. A darby is best used along this area to make sure the surface is flat.

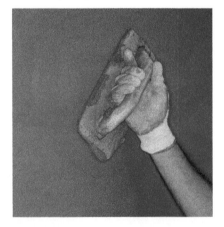

STEP 7 As the applied render surface starts to pull in and set, consolidate the surface with a plastic float to obtain a smooth, plain finish.

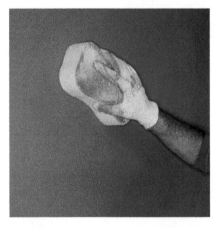

STEP 8 Lightly rub a sponge/sponge float over the face of the finished surface.

▲ Figure 4.43 Sponge float

▲ Figure 4.44 Power float

INDUSTRY TIP

Many plasterers applying modern render systems use a power float for the consolidation and finish of plain-faced renders.

Applying external render to returns using two coat work

Applying render using a scratch and float finish to a return or a pillar can be a difficult technique to learn if traditional handheld methods are used to form the corner. There are several ways to form the corner of the return, which include forming and finishing two angles at the same time (unless using and fixing pre-made angle beads).

Reverse rule/hard angle method

One method of forming the angle is to hold a straight edge along the edge of the return to the required thickness of the scratch coat, checking for plumb with a level. Apply the render to the edge and flatten the surface, then key with a comb scratcher, working away from the edge. Remove the straight edge by tilting one side and repeat the process on the other side. This is known as forming the corner free hand or the 'reverse rule/hard angle' method.

▲ Figure 4.45 Removing the straight edge

The following step-by-step instructions show how to form a return to a pillar (i.e. the top coat finish). This activity needs two people to complete it.

STEP 1 Hold a straight edge to the face of the reveal to a thickness of 10 mm, using a spirit level to ensure it is plumb.

STEP 2 Apply the render mix up to the edge, starting from the top and rule the surface with a darby or small straight edge. Do not work down to the bell cast at this stage.

STEP 3 Once the face has been ruled, remove the straight edge and clean its face.

STEP 4 Place the straight edge on the other face to the same thickness as before.

STEP 5 Apply the render mix up to the edge and rule the surface with a darby or small straight edge. Do not work down to the bell cast at this stage.

STEP 6 Tilt the straight edge and slide away from the angle. This should leave it sharp.

STEP 7 Form the bell cast in the same way, then leave to set before the next step.

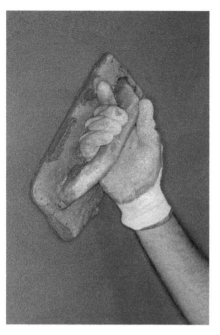

STEP 8 Once the render surface has begun to set and pull in, rub up the face and consolidate the surface with a float.

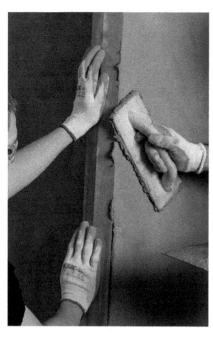

STEP 9 Hold the straight edge to within 1 mm along the edge line and rub up to the edge with the float, consolidating and forming the finish. Again, do not go down to the bell cast until both sides of the return have been finished.

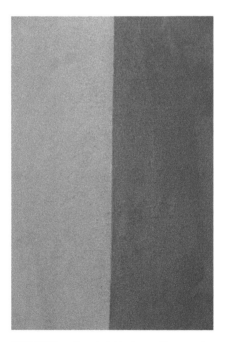

STEP 10 Place the straight edge on the opposite side and repeat the process. Once complete, tilt and remove the straight edge carefully to leave a sharp arris along the edge.

STEP 11 A small plain timber batten can be used to form the edge of the bell, using the same procedure as above.

External angles can also be formed with angle beads, which make this job simple and easy to achieve. These beads can be fixed to the corner, forming and reinforcing a sharp arris that allows you to apply your render and finish flush and flat to the edge of the bead. Once the bead is fixed it will not move, making it easier to apply your render against a solid corner profile.

▲ Figure 4.46 Fixing angle beads to an external corner

All types of bead can be mechanically fixed or direct bond fixed. It is good practice to mechanically fix using fir tree fixings and direct bond fix using wet render product.

- Stop beads are useful profiles that allow the render finish to be stopped against the edge of different surfaces such as face brickwork, timber or plastic cladding. The principle for fixing is the same as for angle beads. Again, once they have been covered with the render material, they cannot be moved or repositioned. Remember to take care when fixing, using a straight edge to make sure the edge of the bead is aligned.
- Expansion beads are fixed to straight joints of brickwork and blockwork. They allow for slight movement, preventing the surface cracking. Two stop beads are used side by side with a gap in between. It is quite common today to have expansion joints in buildings, especially if the building has a steel frame that contracts and expands.

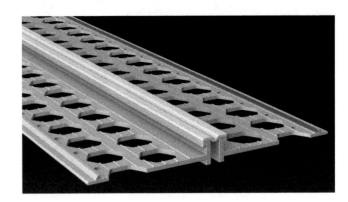

▲ Figure 4.47 A sealed unit expansion/movement bead, which has limited movement potential

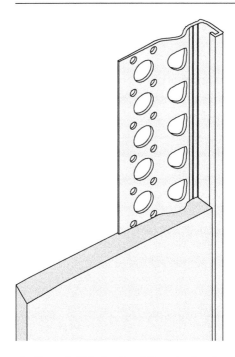

▲ Figure 4.48 A stop bead. Two stop beads can be positioned back to back to form an expansion/movement joint, which has more flexibility for expansion and contraction

External render features

External render features are a good example of how smooth render can be used as a raised feature to produce a classical and elegant look to the wall surface.

Window reveals and heads are rendered smooth, forming a sharp edge for different external renders such as pebble dash and rough cast. Different types of profile can be formed by an experienced renderer to enhance the finish of a rendered substrate: for example, **mullions** around windows, enhanced heads and **cills** (also spelled sills), **key stones**, **quoins** or raised **banding** and **plinths**.

KEY TERMS

Mullion: vertical bar detail around windows.

Cill (sill): horizontal slat detail, forming the base of a window.

Key stone: detail at the apex of a formed arch.

Quoin: detail formed at external corners of a building.

Banding: horizontal detail formed around a building at strategic points.

Plinth: the surface area below the bell cast that runs along the DPC; upstand detail formed at the bottom of a building at ground level.

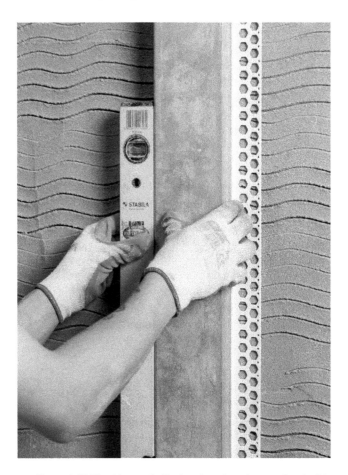

▲ Figure 4.49 Plumbing and aligning the edge of a stop bead with a straight edge and level

▲ Figure 4.50 Window bands and quoins

▲ Figure 4.51 Smooth rendering around a window reveal and soffit

ACTIVITY

Calculate the following areas from the drawing provided:

1 Gable wall to be rendered.
2 Window and door openings.
3 Gable wall without windows/door.
4 Linear length of all beads (stop bead/angle bead/Bell cast bead).

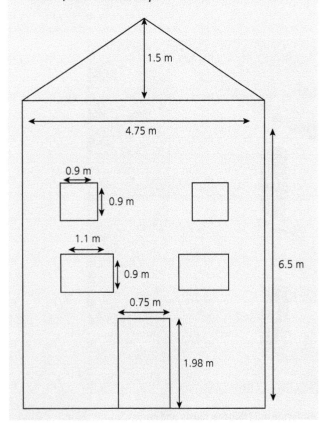

Common faults in external rendering

Faults occur in plastering work due to a number of factors and issues. This table describes some faults that can happen in external rendering.

Fault	Description and consequences
Poorly prepared background	Render applied to background surfaces should be slightly weaker; applying render that is stronger than the background will cause stress, leading the plaster to crack and blow.
	Backgrounds that have poor adhesion and low suction will not bond and the render will become loose in time.
	Render mixes applied over high suction backgrounds will lose their moisture content too quickly, forming a fine crazing surface.
Climate conditions	Hot and cold climate conditions can affect the set of render mixes in different ways.
	Mixing in freezing conditions will affect the strength of the mix and cause the mix to crumble when set.
	Lime blooming is a phenomenon that can affect the early stage of curing of cement render in damp or cold conditions.
	Avoid applying render in direct sunlight as this can draw moisture from the applied render mix, causing it to crumble and become weak.
	Hot conditions can remove the moisture content from the mix; this will affect the curing process and can weaken the mix.
	Rain and frost can cause damage to the surface and strength of the render.
Over-mixing or using contaminated water	Over-mixing rendering materials or using contaminated water affects the setting process and can cause the mix to lose its strength, causing the surface to become powdery.
	Using contaminated water can affect the strength of the render and cause inconsistent colour or colour deterioration.
Incorrect gauging	Incorrect gauging of rendering materials causes mixes to be inconsistent and have different strengths.

KEY TERM

Lime blooming: this happens when lime in the form of calcium hydroxide migrates and forms on the surface as the material dries out. On reaching the surface, this reacts with carbon dioxide in the air and produces a surface deposit of calcium carbonate.

CASE STUDY

Carly works for a local building company as a plastering supervisor. One of her duties was to carry out a rendering survey on a detached bungalow in order for the company to tender for re-rendering.

When Carly arrived on site, she noticed that the plaster was loose and it had severe cracks. It would need to be hacked and re-rendered. When she removed some of the loose plaster, it revealed a lightweight block surface which immediately told her that the background properties were weaker than the render mix. The strength of the mix had caused stress on the background and it had blown from the surface in places. The window reveals had also corroded and there was rust staining on the surface of the render, caused by using the incorrect type of metal bead.

Carly wrote her report and stated that the background would need to have stainless steel EML mechanically fixed to its surface to reinforce it, making it suitable to receive a two coat cement-based render. It would also require either plastic or stainless steel bell beads and angle beads to all openings, including reveals and soffits.

During Carly's visit she was able to determine access for scaffolding and identify clear labour and pedestrian routes to carry out the job safely. She also made enquiries to the client about services and a designated mixing area.

Carly advised the client to use a pre-mixed render finish which comes in a range of colours and would not need painting. This would ensure a low-maintenance solution. It would also benefit the company, as space for mixing was restricted and using pre-mixed render would take less room than mixing a range of traditional render ingredients.

Test your knowledge

1 Why is external render applied to a building?

   A to prevent stains from appearing

   B to prevent paint from flaking

   C to prevent an uneven surface

   D to prevent water ingress

2 To what are dubbing out coats applied?

   A uneven backgrounds

   B flat backgrounds

   C straight surfaces

   D poor surfaces

3 Why is waterproofer added to the mix?

   A to prevent quick curing

   B to prevent fast drying

   C to prevent penetrating damp

   D to prevent moisture rising

4 What can be used to key renders?

   A a comb scratcher

   B a gauging trowel

   C a bucket trowel

   D a mixing tool

5 What are bonding agents used to improve?

   A curing

   B strength

   C adhesion

   D consistency

6 Which one of the following is known as a binder when mixing render?

   A sand

   B cement

   C additive

   D inhibitor

7 What do manufacturers' instructions provide information about?

   A binders

   B aggregates

   C application

   D wastage

8 Complete this sentence: Mixing render with contaminated water will affect the mix's:

   A strength

   B application

   C thickness

   D evenness.

9 Where will you find information about the number of pre-formed beads needed for a rendering contract?

   A schedule

   B data sheet

   C specification

   D bill of quantities

10 When measuring quantities of materials, where will you find the measurements you need?

   A method statement

   B schedule

   C risk assessment

   D drawing

11 Which one of the following tools is used for ruling the surface of external render before it can be finished?

   A boat level

   B adjustable staff

   C straight edge

   D timber batten

12 Complete this sentence: Pre-mixed render is consistent in:

   A strength

   B thickness

   C application

   D suction.

13 Which of the following is used to form external render returns or reveals?

   A angle tool

   B angle guide

   C long timber staff

   D plane timber rule

14 Why would external render material sag after application?

  A It is too thin.

  B It is too thick.

  C It was applied from the top.

  D It was applied from the bottom.

15 Which one of the following floats is used to form plain face rendering?

  A plastic

  B sponge

  C devil

  D setting

16 Why are render surfaces keyed?

  A to prevent suction

  B to allow the next coat to bond

  C to prevent adhesion

  D to improve workability

17 What are external render bell casts used to form?

  A an arris

  B an edge

  C a recess

  D a drip

18 Which of the following is **best** to mix external renders?

  A drill and whisk

  B mechanical drum mixer

  C plunger

  D bucket trowel

19 What is **best** used to gauge render material for consistency and strength?

  A paddle

  B bucket

  C shovel

  D scoop

20 Complete this sentence: External render mixes applied to the background surface **must** be:

  A bulky

  B fatty

  C stronger

  D compatible.

CHAPTER 5

FLOOR SCREED SYSTEMS

INTRODUCTION

Floor screeds are laid to provide a hard, flat, level and durable surface that forms a base for floor coverings such as carpets, vinyl and ceramic tiles. Floors must be level if they are to receive furniture such as kitchen units, tables and chairs. An exception is when preparing a shower area with falls to a gully.

The basic principles for preparing, setting out and laying floor screeds and curing the materials are always similar.

Traditionally screeds were laid by plasterers and if you work in the domestic market for small builders, this is a service that you will generally still provide. However, commercial floor screeding is now usually carried out by specialist floor screeding companies, especially since the introduction of pumped, ready-mixed, easy flow materials. The addition of polypropylene fibres in sand and cement mixes to improve reinforcement and flexural strength is now commonplace, due to the popularity of under floor heating.

By the end of this chapter, you will understand how to:

1 interpret information from drawings and specifications for laying sand and cement screeds
2 select materials and components for laying sand and cement screeds
3 know the difference between the types of screeded floor
4 lay sand and cement screeds.

The table below shows how the main headings in this chapter cover the learning outcomes for each qualification specification.

Chapter section	Level 1 Diploma in Plastering (6708-13)	Level 2 Diploma in Plastering (6708-23) Unit 223	Level 2 Technical Certificate in Plastering (7908-20) Unit 204
Interpret information	n/a	Learning outcomes 1 and 2	Topics 1.1, 1.2, 2.1, 2.2, 2.3
Select materials and components for laying sand and cement screeds	n/a	Learning outcomes 3 and 4	Topics 2.3, 3.1
Types of screeded floor	n/a		Topics 2.1, 2.2
Lay sand and cement screeds	n/a	Learning outcomes 5 and 6	Topics 2.4, 3.2, 3.3

1 INTERPRET INFORMATION

Specifications and drawings

As with all plastering operations, the contract specification is key to producing work to the required standard. This will provide information about the:

- mix ratio
- screed materials to be used
- screed thickness
- insulation thickness
- type of finish
- falls ratio (if the screed is in a shower room or other sloped floor).

▲ Figure 5.1 Floor screeding

The inclusion of information about **insulation** reflects a government focus on saving energy, with Part L of the Building Regulations 2010 in England requiring all properties to supply an **energy certificate**. This has led to increased use of insulation to improve **U-values** in screeded floors. Underfloor heating, which is increasingly popular, also relies on good design and use of insulation.

As well as the contract specification, the contract drawings should also be consulted for information about floor screeding. The main two drawings are:

- a detailed drawing showing how the screed is built up
- a location drawing showing the position of drainage outlets for shower rooms.

INDUSTRY TIP

Good information sources for laying floor screeds are specifications, schedules, manufacturer's technical information (MTI), drawings and building regulations.

KEY TERMS

Insulation: objects or materials used in buildings to improve thermal quality.

Energy certificate: states a property's energy efficiency and recommends how energy can be saved, to save money and be environmentally friendly.

U-values: a measure of heat loss through a building's walls, floors and roof. Higher U-values suggest poor thermal performance. The lower the U-value, the better the building is at retaining heat.

Specification: instructions stating the standards required and practice to be followed for a task, usually BBA-approved and to meet British Standards. It is often an official document from the architect who is overseeing a project.

Schedule: a timetable or sequence of events.

Manufacturer's technical information (MTI): technical information on products for safe use and correct installation.

Drawings: provide a graphic illustration/representation of what is to be built.

Building regulations: rules enforced by the building control department of local councils to ensure all buildings are safe and fit to live and work in. These regulations contain the minimum standards for design, construction and alterations to buildings.

ACTIVITY

1 Search online to find out how U-values are calculated.
2 Work in small teams to create a poster presentation to share your findings.

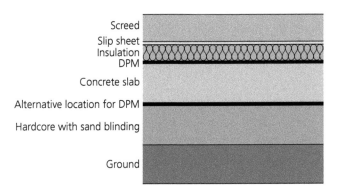

Screed
Slip sheet
Insulation
DPM
Concrete slab
Alternative location for DPM
Hardcore with sand blinding
Ground

▲ Figure 5.2 Example of a detailed drawing

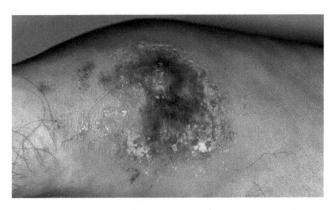

▲ Figure 5.4 Skin burns from cement

ACTIVITY

Carry out an internet search for 'floor screed insulation'. Find the names and websites of two manufacturers that make insulation.

▲ Figure 5.3 Safety glasses

▲ Figure 5.5 Wearing the correct PPE

ACTIVITY

1 Research what is meant by pH of 12–14.
2 Find out what treatment is required if cement burns occur.

HEALTH AND SAFETY

There are many health and safety issues to consider when you are mixing and laying screeds. For example, the dust created by the cement when it is being added to the mix can cause respiratory illness. Measures must be taken either to reduce the levels of dust and/ or to reduce your exposure to the dust, such as:

- mixing in a well-ventilated area (with doors and windows open)
- making good use of LEV
- wearing a fine-particle dust mask or respirator when mixing
- wearing gloves and safety glasses, as well as the standard hi-viz clothing and safety boots.

When the mix is being laid, it no longer poses a significant risk from fine dust particles. However, cement is an irritant and can cause severe chemical

burns. This is made worse by the abrasive (scratching) action of the sand and by kneeling in it as you work. To prevent this from being a problem, wear a long-sleeved, heavy-duty pair of moisture-resistant overalls and a good pair of knee pads.

Wet cement can cause chemical burns on the skin. When water is added to cement, relatively harmless calcium oxide turns into calcium hydroxide which has a pH of 12–14. This can often be slow to injure the affected skin area, meaning that by the time symptoms of burning appear, it is too late and the injury will need treatment immediately.

Laying floor screeds requires you to work while kneeling, bending over and stretching. This can put a lot of strain on your back, so avoid working for long periods without a break.

Mix volume and ratios

Floor screeds are expected to last at least 60 years, so it is important to gauge and mix the floor screeding materials correctly, according to the specification. If you do not do this, there might be cracking or weak spots. To minimise the risk of this happening, a mix ratio is used which stops the mix being too strong or too weak. The two most common mix ratios for a traditional floor screed are 3:1 and 4:1.

- 3:1 – three parts of sharp screeding sand to one part of Ordinary Portland Cement (OPC).
- 4:1 – four parts of sharp screeding sand to one part of OPC.

The materials are mixed by volume to make sure the consistency remains the same throughout the gauge. It is very important to add the right volume of water to your mix, as too much water will weaken the floor screed.

EXAMPLE

If a room measures 5.5 m by 3.5 m and the screed thickness is to be 75 mm with a ratio of 1:4, what is the volume of the materials required?

Step 1

Find out the volume to be filled with screed. Multiply the area of the room by the screed thickness, making sure you are using the same units of measurement.

$$5.5 \times 3.5 \times 0.075 = 1.44\text{ m}^3$$

Step 2

Divide the volume to be filled by the total of both sides of the ratio. The ratio here is 1:4, so 1 + 4 = 5.

$$1.44 \div 5 = 0.288\text{ m}^3$$

Step 3

Multiply each material by the volume it needs to fill. Cement:

$$1 \times 0.288 = 0.288\text{ m}^3$$

Sharp sand:

$$4 \times 0.288 = 1.152\text{ m}^3$$

Step 4

Work out the weight of cement needed. Cement weighs 1280 kg per cubic metre, written as kg/m^3. So multiply the weight per metre cubed by the amount of cement needed in metres cubed.

$$1280 \times 0.288 = 368.64$$

▲ Figure 5.6 Mix ratio of 3:1

Calculating materials

You must know how to calculate the volume of materials to use on a screeding job. Let's look at some examples of how to do this.

Then divide this amount by the weight of one bag of cement (25 kg).

$$368.64 \div 25 = 14.74$$

Round this figure up to the nearest whole number. In this example, the number of 25 kg bags of cement needed is **15**.

Step 5

Next, work out the weight of sand needed. For our calculations we will assume screeding sand weighs 1.6 tonnes per metre cubed, or 1600 kg/m^3. Multiply the weight per metre cubed by the amount of material needed in metres cubed.

$$1600 \times 1.152 = 1843.2\text{ kg}$$

Then divide this amount by the weight of one bag of sand (25 kg).

$$1843.2 \div 25 = 73.728$$

Again, round this figure up to the nearest whole number. In this example, the number of 25 kg bags of sand needed is **74**.

Step 6

Total up your materials. In this example, to lay a screed 75 mm thick with a ratio of 1:4 in a 5.5 m × 3.5 m room, we need 15 bags of cement and 74 bags of sand.

INDUSTRY TIP

When gauging sand by volume, you must consider the sand's water absorption. For example, for domestic work most sand is purchased in jumbo bags, which most people assume to weigh 1 tonne. But on average a jumbo bag weighs approximately 850 kg. To complicate things further, a cubic metre of screeding sand has a dry weight of approximately 1.6 tonnes per cubic metre.

If we ordered the sand in jumbo bags weighing 850 kg, then we would need to divide 1843.2 kg by 850 kg, which is 2.16, rounding up to 3 jumbo bags of screeding sand.

IMPROVE YOUR MATHS

When calculating screeds, you will need to know how to calculate:

- area – the amount of space taken up by a 2D surface
- volume – the amount of space a substance occupies
- linear length – arranged in a straight line
- ratio – the proportion of different materials in a mixture.

2 SELECT MATERIALS AND COMPONENTS

Tools and equipment

To carry out floor screeding successfully, you will need the tools and equipment shown in the table.

Tool	Use
Floor laying trowel	Used to trowel the floor screed smooth. This differs from a plastering trowel as it is made from thicker steel and is at least 450 mm long.
	Some screeding trowels have a pointed end, to allow the plasterer to trowel into the corner of a room.
Float	Plasterers prefer to use a larger float to cover more of the surface area when finishing the work and consolidating the material.
Spirit level	At least two good quality levels are required: one with a length of 600 mm and the other of 1800 mm. These are used to ensure a level surface.
Water level	Used to transfer a **datum point** (see page 204) from one room to another.

Tool	Use
Laser level 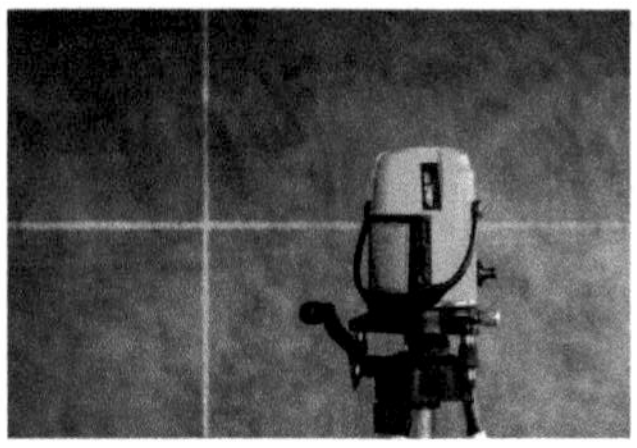	Easy to use. Their accuracy generally increases in line with their cost.
Gauging trowel 	Used to mix small amounts of material and position material into tight corners.
Chalk line 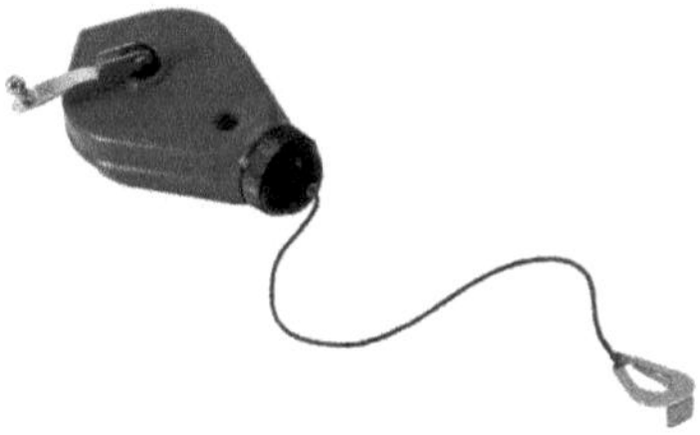	Used to mark out screed lines and **datum lines**.
Box rule 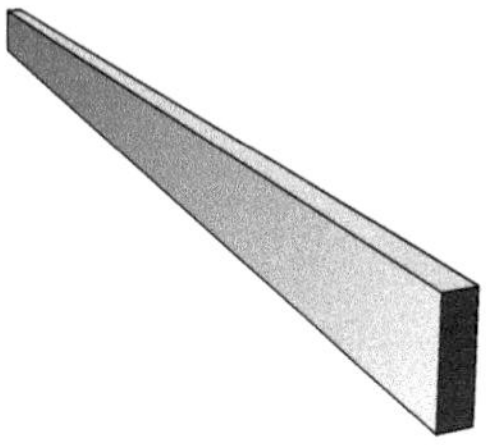	Also known as a flooring rule. Used for ruling in screeds and checking the level, as well as to compact sand and cement down to the screeds. This expels trapped air and compresses the sand and cement, making the screed more solid and helping to prevent weak spots.
Measuring tools 	Tools such as a tape measure are used to measure lengths of timber battens or floor areas.
Square 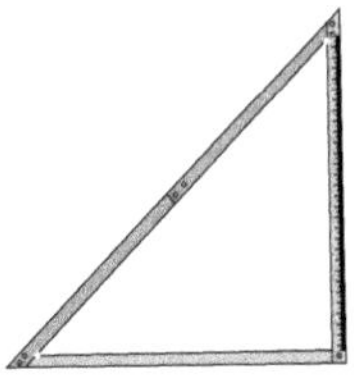	Used to square off frames and walls and for setting the screeds at the datum level.

Tool	Use
Cement mixer 	For domestic work, an electric drum mixer (pictured) is the most popular, but on larger sites a pan mixer might be used. As ready-mixed screed is becoming more popular, this equipment is not always required.
Wheelbarrow 	Narrow wheelbarrows are used for domestic work as they allow you to manoeuvre through doorways.
Large shovel 	Used for mixing and placing mixed material when floor laying.
Buckets	Buckets of various sizes are used to carry water and materials.
Screed rail 	To help keep the floor flat, to form a chequerboard framework or to use when forming falls.

KEY TERM

Datum point/line: a point or line from which measurements are taken to establish the finished floor level; usually about a metre high and running throughout the whole building.

INDUSTRY TIP

The terms 'tools' and 'equipment' are sometimes used interchangeably, but can be used to mean different things. Tools may be handheld but equipment usually refers to larger items. If you are employed, equipment will usually be supplied by your employer.

ACTIVITY

Do some online searches to find out:

- how to use a laser level
- information on datum levels and deviation from datum, using the search term 'Screed Scientist Lexicon'
- in which situations a power float should be used rather than a polyurethane hand float.

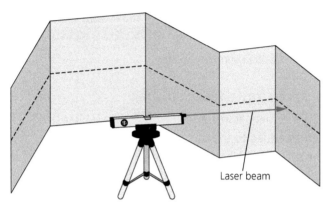

▲ Figure 5.7 Using a laser level

Laser beam

INDUSTRY TIP

A power float is a mechanical tool used instead of a polyurethane hand float to consolidate larger areas of screed work.

▲ Figure 5.8 Power float

Materials

The main two materials used for traditional floor screed mix are:

- Ordinary Portland Cement (OPC)
- sharp screeding sand (washed well-graded sand).

Other materials can also be added in; these are covered later in the chapter.

Ordinary Portland Cement

Cement works by wrapping itself around the **aggregate** (in this case, the sand) during mixing, acting as a binder. Refer back to Chapter 3, page 100, for more information about cement.

OPC is made from limestone and silica. It is produced worldwide, with the industry growing at a rate of 5% per year. It takes a lot of heat to produce a bag of cement, which has implications as governments are trying to reduce CO_2 emissions following the **Kyoto Protocol**.

KEY TERMS

Aggregate: a material made from fragments or particles loosely compacted together. It gives volume, stability, resistance to wear or erosion. Coarse- to medium-grained material used in construction for bulking.

Kyoto Protocol: an agreement between the world's nations to reduce greenhouse gas emissions.

Sharp screeding sand

Coarser than plastering sand, sharp screeding sand has the strength and durability required for a floor screed. The sand provides filler to the mix.

Sand for screeding should be sharp and gritty, containing larger grains of sand to give more durability to the screed. It should also be clean and free from impurities.

▲ Figure 5.9 Sharp screeding sand

When sand is wet it increases in volume, a process called **bulking**. This must be taken into account when gauging materials, to avoid accidentally creating a weaker mix.

KEY TERM

Bulking: the swelling of sand when it is wet, making it heavier.

IMPROVE YOUR MATHS

Remember that 1 tonne = 1000 kg.

ACTIVITY

Find out the cost of 1 tonne of sharp screeding sand by visiting local builders' merchants/suppliers or looking at their websites.

Other materials

Polypropylene

Polypropylene fibres that are 20 mm long can also be added to the mix. They help to reduce cracking and improve strength. As a proportion of the concrete or screed mix, 1 × 100 g bag is used for 1 × 25 kg bag of cement.

In a standard-sized cement mixer (with a 90-litre capacity), one-third of the cement bag is used at a time to create a standard concrete/screed mix, so one-third of the fibre bag's contents would also be added to each mix.

▲ Figure 5.10 Polypropylene fibres

Styrene-butadiene rubber

Like a cement slurry, styrene-butadiene rubber (SBR) can be added to the mix to improve adhesion to the sub-floor. When it is added to a screed mix, the polymers in the liquid bond the aggregate and cement together. This produces high-strength screeds and is useful for carrying out patch repairs.

Hardeners

Hardeners can be added, either during the mixing process or during the **curing** of the floor screed, when the cement is kept moist to allow the screed to harden. They make the surface of the screed more durable and also allow earlier **foot traffic** over its surface.

Waterproofers

These can be added to the mix if the screed is to be laid in a wet area.

Reinforcing mesh

Reinforcing mesh – such as D49 mesh or chicken wire – can be laid midway in the screed to reinforce the screed and help prevent cracking.

KEY TERMS

Curing: allowing the mix to set and reach its full strength.

Foot traffic: people walking or travelling over an area. The term is used when a newly laid floor screed allows someone to walk over the screed without leaving any indentations, usually after about three or four days.

Component materials used for screeding

Pre-mixed screed

Using ready-mixed screed has a few advantages over using a traditional sand and cement site-mixed screed. These include:

- consistent batching of materials
- no storage or mixing areas required on site
- retarders can be added to allow a longer working time
- other additives can be incorporated into the mix, such as polypropylene fibres to reduce cracking and improve **flexural strength**
- increased productivity as site operatives do not have to spend time mixing materials.

KEY TERM

Flexural strength: the ability of a material to bend without cracking or breaking.

INDUSTRY TIP

Screeds are specified by:
- performance – this specifies a quick drying sand and cement mix
- strength – this is specified as a C rating (from C16 to C30), with differently rated mixes being used for various applications
- use and/or wear – this is classified as light or heavy.

▲ Figure 5.11 Pre-mixed screed used for renovations or small projects

Pre-mixed screed can feed into a mixer with a built-in pump, which pumps the mixed screed up many floor heights. This saves production time and operative fatigue.

▲ Figure 5.12 Diesel engine screed pump with hose, tripod, seals and safety clips

IMPROVE YOUR MATHS

Research diesel engine pump screeds.
1. Find out how much they cost to purchase and to hire for one week.
2. Make a table of your findings and report on whether purchase or hire is more cost-effective.

Dry silos

On most medium to large construction sites, dry silos are used to store pre-mixed materials such as mortar, render and floor screed materials. They are connected to the mains water supply, which is regulated to add the right amount of water to mix the materials to the correct consistency. The materials can then be drawn off as needed.

Silos can hold as much as 16 tonnes of material, with some silos having the technology to monitor how full they are and trigger a reorder when the supplies drop to a set level.

▲ Figure 5.13 Dry silos

Expansion strips

To help reduce possible shrinkage and cracking, a large screeded area is laid out in a grid pattern to form **day work joints**. This allows the plasterer to screed the bays in sequence, rather than trying to screed the whole area in one go. Each bay is approximately 14 m². Expansion strips are placed and levelled into each bay at its joins.

An added advantage of using an expansion strip is that it can be used as a screeding point to rule from. The current guidelines ((BSI) **BS5385**) suggest that expansion strips should placed every 8 m.

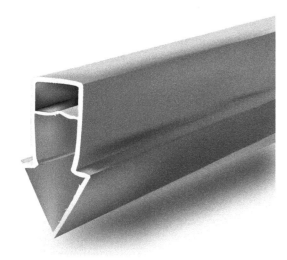

▲ Figure 5.14 Day expansion strip

Damp proof membrane

Commonly abbreviated to DPM, a damp proof membrane is a thin, plastic, sheet-like layer. It is laid in the sub-floor to prevent moisture rising up from the ground and into the screed. The membrane is **lapped** up the wall level with the damp proof course (DPC); see Chapter 1 for more information.

Materials such as lining paper for decorating and polythene for flooring are sold according to their thickness. The current building regulations state that polythene used for a DPM must have a thickness of at least 1200 gauge.

▲ Figure 5.15 Lapped DPM

Building sand

Soft building sand can be laid and compacted to an average thickness of 50 mm over the sub-base concrete; the DPM is then laid on top of the building sand. The sand minimises the risk of concrete aggregate puncturing the DPM.

This process is known as sand blinding.

Liquid damp proof membrane

Liquid DPMs are sometimes used in refurbishment work where an existing floor screed is failing. The old floor screed is removed or cleaned to receive a coat of bitumen liquid that is allowed to dry overnight. Once dry, a second coat of bitumen is brushed over the first coat.

While the second coat of bitumen is still tacky, clean sharp sand is scattered over it and allowed to dry. The new screed can then be laid on top, to a minimum thickness of 50 mm.

Also available is a two-pack epoxy liquid-applied DPM, which is solvent-free and therefore low odour. It is used to provide protection for moisture-sensitive floor finishes when the DPM is unsatisfactory and might fail. The pack is manufactured in two contrasting colours to help identify which areas have been coated.

Rigid insulation

Rigid insulation is used under the floor screed or under the sub-base (see Chapter 1) to help reduce heat loss through the floor.

Rigid insulation is available in various thicknesses, to allow for the horizontal and vertical positioning of the insulation. The insulation is very durable and offers good **compressive strength**. On this type of board, the minimum thickness of the screed should be 65 mm depth.

ACTIVITY

1 With a partner or in groups, discuss the steps and processes for creating a liquid DPM.
2 Search online using the term 'floor screed insulation board' and find out about Kingspan Thermafloor TF70.

KEY TERM

Compressive strength: the ability of a material (for example, insulation) to take heavy loads, such as furniture and people, without denting or going out of shape.

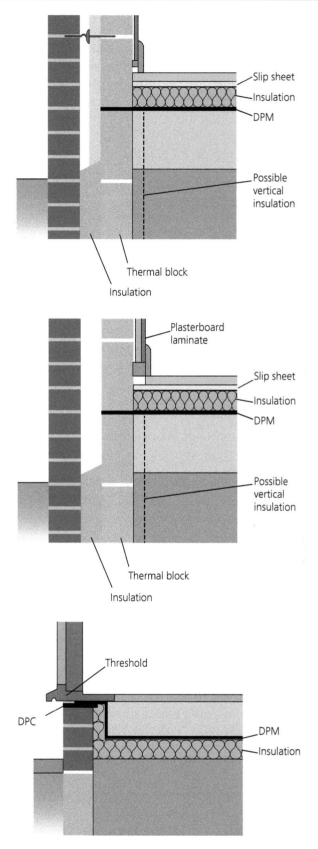

▲ Figure 5.16 Insulation at various positions

Slip sheet layer

A slip sheet is laid between the insulation and the floor screed, made from 1200 gauge polythene sheet. It should be lapped 150 mm up the sides of the walls to form a vapour control.

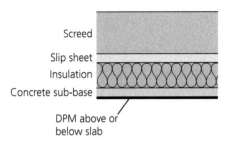

Screed
Slip sheet
Insulation
Concrete sub-base

DPM above or below slab

▲ Figure 5.17 Slip sheet

Self-levelling compound (latex)

Self-levelling compounds are used to compensate for irregular floor screeds, especially on refurbishment work.

First prepare the floor which is receiving the compound by removing loose particles and dust. Then apply a waterproof PVA (if necessary, following the manufacturer's technical information) before self-levelling, to control suction and improve the self-levelling compound's adhesion.

All self-levelling compounds are cement-based. The cost of self-levelling compounds varies, depending on the quality of the aggregates and their flow rate (ability to self-level). Better quality self-levelling compounds contain latex, which is either incorporated in the dry mixture or added separately during the mixing process. Self-levelling compounds that contain latex flow better and provide a smoother finish if laid correctly.

The method is the same, whether or not compounds contain latex:

- Mix the compounds in a bucket with a mixing whisk until they have a creamy consistency.
- Lay immediately with a plastering trowel to the floor, to a thickness of between 2 and 5 mm.

Self-levelling compounds are available in slow set and fast set varieties. The fast set variety can receive foot traffic after 20 minutes. It is important to lay the self-levelling compound reasonably quickly, to allow the materials to flow and self-level.

Free flow liquid screeds

Free flow liquid screeds use modern laying techniques and materials. Calcium sulphate, sand, water and additives are mixed to produce a free flow self-levelling screed.

- Instead of marking out the screed using dots or battens, **screed levelling tripods** are used, levelled from a datum line.
- The screed is poured over the area through a hose, up to the underside of the tripods.
- A **dappling bar** is then gently moved across the floor to level the floor screed.

This type of screed must be laid over an insulated sub-floor, as moisture can affect its durability. It is a quick process which makes it ideal for larger floor areas.

KEY TERMS

Screed levelling tripod: used to make sure the floor is laid to the correct level.

Dappling bar: used to help level and finish the liquid screed.

ACTIVITY

Find a video online of pump-applied floor screed being laid. One example is Weberfloor 4310 Fibre Flow-Weber Flooring Systems.

▲ Figure 5.18 Dappling bar

▲ Figure 5.19 Free flow screed with tripod

▲ Figure 5.20 Screed levelling tripod

③ TYPES OF SCREEDED FLOOR

There are four main types of sand and cement screed. These are:

- monolithic
- bonded
- unbonded
- floating

Monolithic

When something is described as being monolithic, it means it is made up of a single, unbroken mass. When a monolithic floor screed is laid, it is laid within three hours of the concrete slab substrate being poured, that is while the sub-base concrete is still **green**. By then, the concrete has had its initial set but not its final set. This means the screed which is laid on top chemically bonds with the concrete, as they both set and dry.

The recommended thickness of a traditional sand and cement screed without fibres is 20 mm, but can be from 12 to 25 mm depending on the levels, high points and low points of a floor. With added fibres, the thickness can be reduced to 10–15 mm.

This type of screed is often found in commercial settings, such as factories and warehouses.

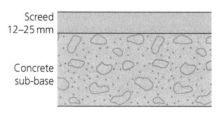

▲ Figure 5.21 Monolithic screed

KEY TERM

Green: describes a material such as concrete/screed/plaster that has not fully set and is still soft.

Bonded

As its name suggests, this type of floor screed is bonded to the substrate. The screed is laid when the sub-base concrete has hardened and been **scabbled**, so that it is left with a **tamped finish** that provides a key for the floor screed.

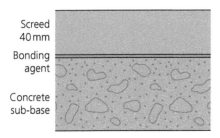

▲ Figure 5.22 Bonded screed

Bonded screed is also known as separate. Before the screed is laid, the sub-base concrete should be soaked with water, ideally overnight, to minimise the risk of bond failure and edge curling. To improve adhesion, a slurry of cement and SBR can be mixed together to a ratio of 1:1 and then brushed over the concrete, and the screed should be laid while the slurry is wet.

Pre-formed concrete slabs and concrete block floors are also suitable substrates for this type of screed. The minimum thickness for this floor is 40 mm.

Unbonded

This type of screed is laid directly on top of either heavy waterproof building paper or 1200 gauge polythene sheeting (DPM) with at least 100 mm laps up the outside walls. The only preparation required is to sweep the substrate clean of any debris that might puncture the DPM.

The screed is laid to a minimum thickness of 50 mm, but this increases to 65 mm if it is laid over an insulated floor containing heating pipes.

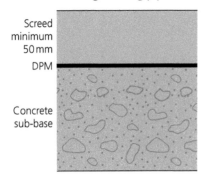

▲ Figure 5.23 Unbonded screed

Floating floor

The floating floor method, also known as unbonded screed, is becoming more common. This type of screed is laid on rigid **extruded polystyrene** insulation to provide an insulated floor area.

All walls and pillars must be lined with **edging foam** or 20 mm insulation, to protect against shrinkage cracking.

This system is the most commonly used in modern house building, especially for floors with underfloor heating systems.

- Floating floor screeds must be laid to a minimum thickness of 65 mm or 75 mm if heating pipes are contained in domestic flooring.
- Commercial floating floor screeds are laid to a minimum thickness of 75 mm and no greater than 100 mm.

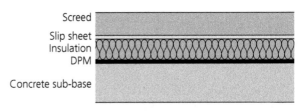

▲ Figure 5.24 Floating screed

KEY TERMS

Scabbling: removing the surface finish by mechanical means, producing a suitable key.

Tamped finish: where the screed has been compacted and consolidated to push coarse aggregate below the screed surface.

Extruded polystyrene: this is formed by heating polystyrene crystals to high temperatures, along with other additives, and forcing the mixture through a die (which is like a mould). The result is a denser material than expanded foam.

Edging foam: this measures 8 mm × 150 mm × 50 m and is used to line the perimeter of each room where either a high floor build-up or a screed layer is required. The foam is used to butt all floor layers to reduce the effect of impacts transferring into the adjacent walls.

▲ Figure 5.25 Edging foam

ACTIVITY

1 In a small area, practise laying a bonded floor, an unbonded floor and a floating floor.
2 Discuss the differences between them.

Floor screed tolerances

All screeded floors should be level and flat. However, it is good industry practice for the quality of the finished floor to match what is written in the specification, which will be based on the British Standard **BS8204-1:2003**.

To check how level a screeded floor is, lay a 3 m rule across the surface of the screed. Any gaps or hollows showing underneath the rule must be within the tolerances shown in this table.

Class	Tolerance for gaps/hollows showing under a 3 m rule
SR1 (the highest specified floor screed finish)	3 mm
SR2	5 mm
SR3 (the lowest specified floor screed finish)	10 mm

INDUSTRY TIP

SR means surface regularity. It is classified as:
1 high standard
2 normal standard
3 utility standard.

ACTIVITY

1 Draw a cross-section of a monolithic floor, a bonded floor and an unbonded floor on DPM.
2 Use the correct symbols for each material found in the floors' construction. See Chapter 1 (page 35, Figure 1.79) for more information on the symbols to use.

Screed sub-bases

Floor screeds are laid onto the following sub-bases:
- green concrete base (when laying a monolithic screed)
- set concrete base (when laying a bonded screed)
- **beam and block** (when laying a bonded screed)
- beam and block overlaid with rigid insulation (when laying a floating screed)
- set concrete base overlaid with polythene membrane (DPM) (when laying an unbonded screed)
- set concrete overlaid with rigid insulation (when laying a floating screed).

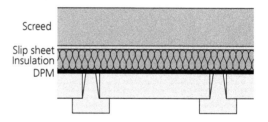

▲ Figure 5.26 Beam and block

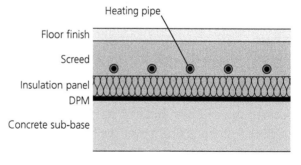

▲ Figure 5.27 Underfloor heating screed

KEY TERM

Beam and block: floors made up of standard building blocks laid between pre-stressed concrete beams.

213

4 LAY SAND AND CEMENT SCREEDS

Preparing floors

Traditional screeded floors are laid with a **semi-dry** mix. Mixing the screeding sand and cement in this way allows you to rule off, compact and consolidate as the work proceeds. Preparation will depend on the type of sub-base.

Consolidating and compacting a screed helps it to bond and stops it breaking up. To do this, tap the loose screed with a trowel or aluminium box rule, to compact it before ruling off.

Providing a mechanical key

When screeding directly to a concrete sub-base that has been allowed to harden, i.e. for a bonded screed, prepare the base as follows.

▲ Figure 5.28 Providing a mechanical key

1 Provide a mechanical key if screeding directly to the sub-base.
2 Sweep and remove all debris and any signs of **laitance**.
3 Soak the concrete sub-base overnight with water. Remove surplus water the next day with a broom.
4 Mix a 1:1 mix of water and PVA to 50% cement to make a slurry and brush over the concrete sub-base.
5 Lay the floor screed while the cement slurry is tacky.

KEY TERMS

Semi-dry: the mix consistency of a traditional sand and cement screed.

Laitance: a layer of weak cement that can affect the strength of the floor screed if not removed.

HEALTH AND SAFETY

- When brushing a dusty floor, wear a dust mask and spray a light water mist over the floor to help keep the dust down.
- Incorrect disposal of waste such as dust could result in the blocking and contamination of drains. Remove waste in line with RAMS.

INDUSTRY TIP

Methods for creating mechanical key:

1 hacking laitance
2 scabbling smooth surfaces
3 brushing newly laid concrete surface.

Methods for creating chemical key:

1 primers
2 cement grout
3 bonding agents
4 bitumen.

▲ Figure 5.29 Sweeping the floor

Screed mixing methods

It is important always to gauge materials by either weight or volume, using a gauge box or buckets. Never gauge materials using a shovel, as this will produce an inconsistent mix. Always read the specification to ensure the materials are mixed to the correct mix ratio.

Mixing by hand

For smaller quantities, such as for repair work, the sand and cement can easily be mixed by hand. The materials should be proportioned by weight or volume following this sequence.

STEP 1 Gauge the materials with buckets.

STEP 2 Place the materials in a single pile.

STEP 3 Mix the materials dry (i.e. without adding water).

STEP 4 Pour a small amount of water into the centre and gently turn the sand and cement into the water using a shovel. Continue until all of the sand and cement is damp. Turn over the material at least three times until thoroughly mixed.

STEP 5 This is a semi-dry mix. It is important not to add too much water when mixing: you should be able to clench a ball of sand and cement in your hand without any water squeezing through your fingers. This is often referred to as 'the snowball test' – the mix should be a similar consistency to a snowball.

INDUSTRY TIP

Using a semi-dry mix allows you to compact the sand and cement screed when ruling off and reduces the risk of laitance forming.

Mixing by cement mixer

This is where the materials are put in a cement mixer and tumbled until mixed.

▲ Figure 5.30 Mixing by cement mixer

Take care when adding water, as the tumbling action of the materials being mixed together can cause balls of compacted materials to form. Always ensure there are no unmixed materials stuck at the back of the mixer.

The following steps explain how to mix with a cement mixer.

1 If using an electric cement mixer, check that it has a current Portable Appliance Test (PAT) label (see page 69) and is safe to use.
2 Make sure the work area is uncluttered and all materials are close to hand with a supply of clean water. Check that the mixer is on a firm standing.
3 If using an additive, add it to a large drum of water to ensure consistency throughout the mix. Use water only from this drum for mixing.
4 Add a small amount of water into the mixer drum.
5 Add fibres, if you intend to use them.
6 Add about half of the sand that you intend to use.
7 Add about half of the cement that you intend to use.
8 Add a little more water, followed by the remaining sand and cement.
9 Mix for no longer than five minutes – three minutes is recommended.

10 Pour the mixed material onto a spot board or into a wheelbarrow.
11 Wash out the mixer and clean the equipment.

HEALTH AND SAFETY

There is no set frequency for PAT testing. However, if equipment is used often, it should be tested at regular intervals by a competent person to make sure that it is safe.

INDUSTRY TIP

Do not mix more materials than you can use within an hour.

Ready-mixed screed

Using ready-mixed screed has advantages compared with mixing the screed yourself.

- There is no need to store materials as the screed can be laid the same day it is delivered.
- Additives that delay the setting time can be added by the manufacturer, allowing more time to lay larger areas.
- Fibres and hardeners can be added by the manufacturer to help reduce cracking.
- The correct consistency of mixed screed material is guaranteed, as the supplier will mix and deliver the screed to the screeder's specified ratio or the architect's design specification.
- There is less wastage – you only need to order enough materials for the job.
- As bagged materials do not need to be stored on site, there is a reduced risk of theft.

INDUSTRY TIP

Ready-mixed materials are most commonly used because:
- there is a guarantee of consistent mix and setting times
- they save time and labour, as no mixing on site is required.

ACTIVITY

1 Look at the ready-mixed section of these websites:
- https://readymixedconcrete.com/calculator/
- https://source4me.co.uk/calculate_readymixed_concrete.php
2 Use your own figures to arrive at an estimate to lay the screed in a room in your home.

Using and laying ready-mixed materials

Ready-mixed screed can be laid using battens or from screed dots. Preparation is the same as for a traditional sand and cement screed.

To be effective when using ready-mixed screed, it is important to set up your working area correctly. For example, when working on a domestic extension, lay out polythene sheets as near to the delivery point as possible so the material can be stored there, with the sheeting preventing contamination of the delivered materials.

On larger projects, ready-mixed screed can be pumped as far as 60 m. Some delivery vehicles can pump as much as 20 tonnes of ready-mixed screed.

IMPROVE YOUR MATHS

If a room measures 5.5 m × 3.5 m and the required screed thickness is 75 mm, how much ready-mixed material is required, not allowing for wastage?

Make sure you convert the measurements so that you are working in the same units. You will need the formula:

$$volume = length \times width \times height$$

Effects of poor workmanship

Cause	Effect
Lack of preparation of sub-base	Cracking and lifting of screed
Too little water	Materials difficult to lay and compact
Too much cement	Surface cracking
Too little cement	Weak screed, prone to wear and tear
Too much water	Weakens screed, causes cement to float to the top of the screed, screed shrinks
Inconsistent mixing	Leaves patches of sand or cement
Over-trowelling or trowelling too soon	Attracts water to the surface with the cement, leaves a film of cement on the surface of the screed
Poor curing	The screed dries out too quickly which weakens it; curing can occur around the edges of the screed
Poor **compaction**	Produces small pockets of air that weaken the floor

KEY TERM

Compaction: consolidation of the sand and cement screed by tamping the screed with a box rule and floor laying trowel. This strengthens the floor screed.

ACTIVITY

Answer these questions, referring to the task above:

1 If the screed is to receive ceramic floor tiles, specify the type of floor finish that will be necessary.
2 Outline how you will promote the curing of the floor once it has been laid.

Floor screed finishes

The purpose of a floor screed is to provide a smooth, level surface that can withstand loads and foot or wheel traffic. There are different ways in which a floor screed can be finished.

Type of finish	Explanation
Trowel finish	This finish is achieved using a large steel trowel for smaller domestic floors or a power float for larger floors. Trowel-finished floors usually receive vinyl tiles/sheets or carpet.
Float finish	This finish is achieved using a large plain float. The float leaves a coarser texture which provides a key for tiling adhesive when laying ceramic floor tiles.
Latex finish	This finish is achieved using a levelling compound, which is trowelled over an existing floor with a steel trowel to a thickness of about 3–5 mm.
Free flow screed finish	This finish is achieved using a dappling bar and is mainly used on larger commercial properties.

Methods of laying a floor screed

The first task when laying a floor screed is to determine the levels to work from. These are taken from a fixed point such as:

- a concrete door step
- the bottom of a door frame for domestic work
- a datum line.

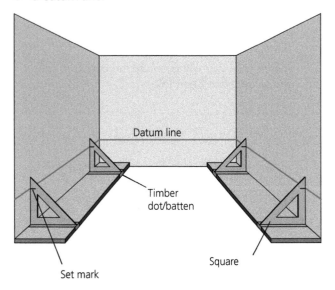

▲ Figure 5.31 Laying to a datum line using a square

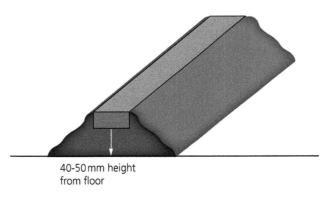

▲ Figure 5.32 A dot bedded in the screeding material

A datum line should always be used for commercial work. Always check that the datum point allows for the floor finish to be laid to the correct finished floor level (FFL).

Once the finished floor level has been determined, timber battens or dots are bedded in, using the screeding material, around the perimeter (outside edge)

of the room and at intermediate positions (points in between) to allow ruling off. When you have levelled the battens or dots and they can stand the pressure of ruling off, screeding can begin.

- If the timber batten method is used, the plasterer/screeder starts to fill in between the timber battens at the furthest point away from the exit point, ruling and tamping the screed materials as the work progresses. The timber battens are usually about 1.5 m long; this allows the plasterer/screeder to remove the battens as they proceed, fill in with floor screed where the batten was positioned and then finish the work with a steel or plastic float.
- Alternatively, if dots are used, screeding materials are laid between the dots and ruled off using a floor rule. As soon as the screed is firm, the timber dots are removed. The plasterer/screeder will start to fill between the wet screeds, ruling and tamping as the work proceeds, finishing the floor with a trowel or steel float. If the floor screed material is initially too wet to finish with a trowel or float, this can be done later in the day when the floor is able to take light foot traffic.

> **INDUSTRY TIP**
>
> To make the screed non-slip, carborundum dust can be sprinkled over the surface and trowelled in at the final trowelling stage. For non-slip wet areas, a batten roller is used to produce a dimpled effect.

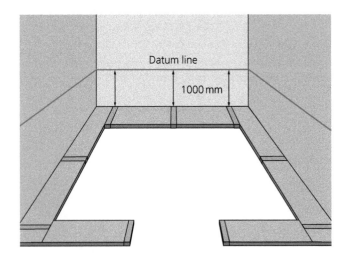

▲ Figure 5.33 Set the dots and form a perimeter screed

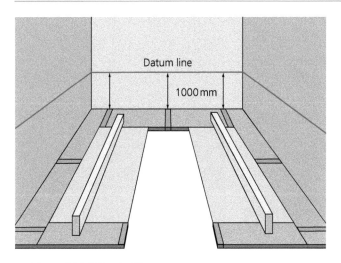

▲ Figure 5.34 Fill both sides

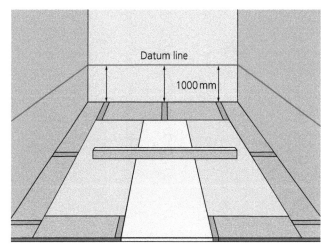

▲ Figure 5.35 Fill middle and work out of the room

The following step-by-step instructions show you how to lay a floor to a datum line using dots.

STEP 1 Damp down the floor.

STEP 2 Sweep the floor.

STEP 3 Set out the dots. The first dot should be set in the corner furthest from the door and approximately 300 mm from the end of the wall which has the room's longest length. Use a tape measure to make sure the dot is the required distance below the datum.

STEP 4 Set further dots as required. Make sure that each dot is the same distance below the datum line. Level the dots across.

STEP 5 Form screeds between the dots. Ensure they are in line and flush with the top of the dots and are level along their length.

STEP 6 Rule in the floor to the dots with a box rule.

STEP 7 Before you lay beyond what you can comfortably reach without over-stretching, remove the dots, fill their holes and float the surface of the floor, filling any misses and hollows, etc. Check the floor with a box rule.

STEP 8 Repeat Steps 5–7 along the opposite wall to form another screed.

STEP 9 You should now have two screeds ruled in to dots and levelled in from the datum.

STEP 10 Fill in between the screeds. Starting at the back wall of the room, empty a wheelbarrow of mix between the screeds. This will need to be compacted down as firmly as possible, then ruled off flush with the screeds.

STEP 11 Follow the same process for the back screed, applying screed and compacting the mix.

STEP 12 Rule in the screed with a box rule and float.

STEP 13 Use a trowel to smooth the screeds.

STEP 14 Continue in this manner, compacting, ruling and floating, working methodically towards the door.

STEP 15 Complete the floor with a trowel.

STEP 16 Carry out a final check for level.

Laying screed to a larger area

Larger areas of floor screed require a different laying technique to reduce cracking and to allow finishing of the floor. Screed rails are levelled through to form a chequerboard framework: this allows the floor screeder to lay alternate bays. The screed rails are then removed and expansion strips are placed in the gaps to reduce cracking.

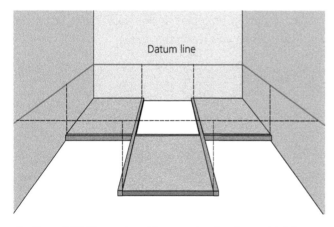

▲ Figure 5.36 A chequerboard framework

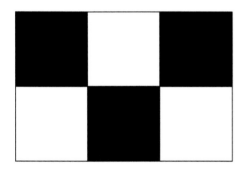

▲ Figure 5.37 Chequerboard framework using day work joints or timber battens

Screed to a fall

All of the basic principles for setting out a sand and cement screed can be applied to laying a screed to a fall (a sloping floor). The purpose of laying screed to a fall is to enable water to drain away to a sunken drainage outlet. This type of floor screed can be found in food preparation areas where washing down is required, in kennels and in walk-in shower/wet rooms.

▲ Figure 5.38 Drainage outlet

IMPROVE YOUR MATHS

A factory floor is to be laid with a constant fall of 1 : 100 throughout its 15 m length from a level base. The minimum thickness is to be 50 mm at the drainage point.

Calculate the maximum thickness of the floor.

Make sure you convert the measurements so you are working in the same units.

Setting out screed to a fall

As with a flat floor screed, the finished floor level is established from a datum point or line. The next stage is to check the working drawing to find the ratio of the fall.

For example, the drawing might state the fall is 1 in 100 or 1 : 100. This means that for every 100 mm in length, the screed will slope down 1 mm. So, if the floor screed was 10 m long to an outlet, the fall (slope) would be 10 cm.

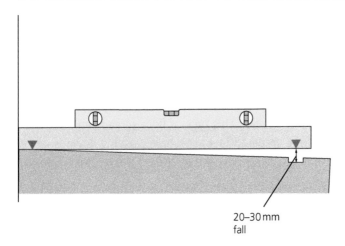

▲ Figure 5.39 Setting out to falls

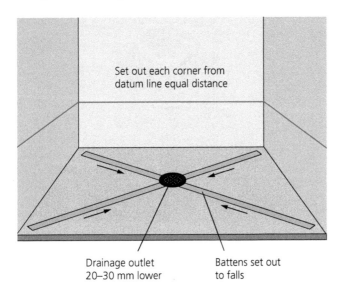

Set out each corner from datum line equal distance

Drainage outlet 20–30 mm lower

Battens set out to falls

▲ Figure 5.40 Falling to a drainage grid

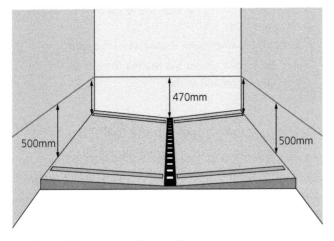

470mm

500mm

500mm

▲ Figure 5.41 Laying a fall to a gully

Laying screed to a fall

When laying a screed to a fall, you have to work from two datum points:

- a horizontal datum line or point to produce the flat part of the floor
- the sunken gully to form the slope that allows the water to drain away.

▲ Figure 5.42 This picture shows forming falls to a grid gully, mostly used in public leisure facilities which cater for multiple showers

Before starting work, check with your line manager or the client that the sunken gully is connected and functioning correctly. As with a flat floor screed, you can use screed rails or form wet screeds to establish the slope, working from the gully to the flat screed area. Depending on the design of the gully, the finished floor level will be flush with the gully's rim or just below it; this is to allow for fixing of ceramic or vinyl sheeting.

Always check the drawings and specification, as no two floor screeds will be to the same design. Working angled screeds is a difficult job. You will need shorter feather edges or box rules as the floor screed becomes narrower when you reach the gully.

▲ Figure 5.43 Using battens to lay a screed to a fall

The following steps show you how to lay screeds to a fall.

STEP 1 Set up the perimeter level, which is always flat. Place a tile in each corner.

STEP 2 Level the corners across.

STEP 3 Transfer the level diagonally.

STEP 4 Fill in the perimeter.

STEP 5 Rule off with a box rule.

STEP 6 Compact the screeds to create the finished floor level.

STEP 7 Lay battens (or form freehand) towards the gully.

STEP 8 The gully will be lower than the screeds.

STEP 9 Rule off the timber battens and compact.

STEP 10 Form the other side of the gully with battens or freehand.

STEP 11 Compact.

STEP 12 Remove the tiles and battens and fill in.

STEP 13 Trowel or float finish.

STEP 14 The finished fall.

Curing and drying out

Curing means keeping cement moist to allow the screed to fully harden. This is different from drying. Curing is an essential part of floor screeding as it allows the cement to reach full strength – the cement should not be allowed to dry too quickly.

There are two options for curing:

1 Keep the floor screed damp for about seven days by covering it with plastic sheeting or lightly spraying it repeatedly with water. After seven days, remove the plastic sheeting.
2 Spray a chemical curing agent onto the floor screed. After a few days, the chemical agent breaks down and curing is complete.

Whichever method you use, it is important to allow enough time. Artificial heat should not be used to speed up drying as the screed needs to be allowed to dry naturally. As a rule of thumb, allow one day of curing for every 1 mm depth of floor screed. It might therefore take a long time.

To test: leave a glass beaker upside down on the floor screed overnight. If condensation appears in the glass in the morning, the floor screed is still drying out.

If any cracks appear during the curing period, there is a simple test for lack of bond. Tap with a hammer either side of the cracked area and listen for a hollow ringing sound. If you hear that noise, it indicates delamination of screed to the substrate base. However, remember that unbonded and floating floors will always sound hollow when tapped.

It is important to protect the working area from foot traffic for several days while setting and curing take place. If after a couple of days some foot traffic needs to access the screeded area, always put timber sheets or scaffold boards down to take foot traffic. If any damage occurs, it is important to do repairs correctly (particularly when preparing and keying the background), replacing damaged sections and leaving a seamless finish.

A 75 mm screed could take three months to dry completely.

CASE STUDY

Ayna and Jordan have been awarded a 12-month contract with an underfloor heating specialist on a sub-contract, labour-only basis.

They received a phone call from the main contractor, giving them details about a job in Leeds. It meant an early start for Ayna and Jordan as they had to travel about 50 miles to get to the job and the ready mix was due to be delivered between 8.30 a.m. and 1.00 p.m.

When they arrived at 8.15 a.m., the plumber was just carrying out a pressure test to ensure there were no leaks in the plumbing system. Where the screed was to be laid, the insulation and the slip layer were already in place, including the perimeter insulation to help prevent cold bridging and allow for expansion and contraction as the floor screed heats up and cools down.

Ayna laid a sheet of polythene on the drive to receive the ready mix, while Jordan fetched some battens to lay over the water pipes to prevent damage to the heating system. Jordan located the datum and marks around the room where the finished floor level should be: he had to make an allowance as the floor was to receive ceramic floor tiling over the screed.

The ready mix arrived at 9.30 a.m. As specified, it included a retarder to allow Jordan more flexibility and polypropylene fibres to improve the flexural strength of the floor: this is essential to help prevent the floor from crazing as the screed heats up and cools down.

Ayna decided she would barrow the materials to Jordan so he could lay the screed to the minimum 65 mm depth as recommended in the British Standards. Jordan started at the furthest point away from the door opening as he didn't want to trap himself in the room. Jordan is an experienced plasterer and floor screeder, so he laid the screeds to work from without using battens; he worked around the perimeter of the room and laid a few intermediate screeds to create bays of workable sizes.

With the levelling screeds laid, he started to fill in between them, ruling and tamping with a box rule to consolidate the ready-mixed sand and cement. Checking as he worked that the floor screed was flat and level, Jordan used a steel trowel to lay the floor, followed by a large plastic float to flatten and provide a textured surface ready to receive the ceramic floor tiles. Ayna waited for Jordan's instructions before dropping the ready mix onto the insulation as it was important that Jordan had control of the laying process.

As Jordan neared the end of laying the screed, Ayna started to clean and tidy the working area. With the floor screed laid, Ayna sprayed it with a curing mixture to make sure the chemical reaction took place: if a floor screed dries out too quickly it can become weaker. Polythene can be used to cure a floor screed but it can only be placed when the screed can take light foot traffic, and by that time Ayna and Jordan would have started a new job.

Test your knowledge

1 A monolithic floor screed must be laid within how many hours of the concrete being placed?

A 3

B 6

C 9

D 12

2 Which one of these is an advantage of using a ready-mixed screed material?

A easier to lay

B cheaper to buy

C consistent mix quality

D consistency can be adjusted on site

3 How much ready mix is required for a floor measuring 6 m by 4 m with a screed depth of 65 mm?

A 1560 m³

B 0.156 m³

C 15.60 m³

D 1.56 m³

4 Which of these is appropriate for floor screed battens?

A They should be left in the screed when the floor is completed.

B They should be removed as work proceeds.

C They should be removed after the floor is completed.

D They can be left in or taken out.

5 Which of these injuries can be caused by exposure to cement?

A skin burns

B dry skin

C cracked nails

D sore knees

6 Which resource should be used to transfer a datum point to another room?

A boat level

B water level

C feather edge

D measuring tape

7 What is laitance?

A a cement layer on top of the finished screed

B an additive to promote curing

C an additive to promote hardness

D a type of floor finish

8 Why is curing necessary?

A for quick drying of the screed

B to aid compaction of the screed

C to obtain an even float finish

D for hardening of the cement

9 What is the **minimum** thickness of a floor screed over insulation?

A 85 mm

B 75 mm

C 65 mm

D 55 mm

10 What is the purpose of perimeter insulation around a room?

A to prevent cold bridging

B to give an edge to work to

C to allow for dry lining

D to keep the floor warm

11 A floor screed finished with a trowel is suitable for which type of floor covering?

A wood block floor

B ceramic floor tiles

C underlay and carpet

D vinyl sheeting/tiles

12 What is self-levelling latex screed used for?

A to cover up poor workmanship

B to level an uneven floor

C to give a trowel finish

D to cover up cracked floors

13 Which one of the following is the **minimum** thickness of floor screed if laid over an insulated floor with heating pipes?

A 75 mm

B 65 mm

C 55 mm

D 50 mm

14 What can be added to increase the strength of a screeded floor?

  A hessian

  b rapid hardening cement

  C polypropylene fibres

  D masonry cement

15 What can poor curing cause a floor screed to do?

  A dry out too quickly

  B weaken around its edges

  C weaken in its centre

  D dry out too slowly

16 What might happen if there is too much cement in a screed mix?

  A The screed will shrink and crack.

  B The screed will have good adhesion.

  C The screed will set quickly.

  D The screed will be weaker than normal.

17 What is the most common mix ratio for a floor screed?

  A 1 sand to 1 cement

  B 2 sand to 1 cement

  C 4 sand to 1 cement

  D 5 sand to 1 cement

18 Floor levelling compound should be mixed to what sort of consistency?

  A thick

  B watery

  C creamy

  D weak

19 What is used to lay floor levelling compound?

  A steel trowel

  B plastic float

  C gauging trowel

  D bucket trowel

20 A monolithic screeded floor has a fall of 1:100 throughout its length of 15 m from a level base. With a minimum thickness of 25 mm, what is the **maximum** thickness of the floor?

  A 125 mm

  B 150 mm

  C 175 mm

  D 200 mm

CHAPTER 6

PRODUCING, FIXING AND FINISHING PLAIN PLASTER MOULDINGS TO MATCH EXISTING

INTRODUCTION

To be a fibrous plasterer you need to develop lots of new skills. This chapter will help you acquire and develop them. It explains the process from the design and planning stage to constructing running moulds, producing reverse moulds and casting from reverse moulds.

By reading this chapter you will gain knowledge of:

1 what fibrous plaster mouldings are and how to interpret information about them
2 how to select materials, components and equipment for producing fibrous plaster moulds
3 how to produce plaster reverse moulds
4 how to cast from a reverse mould
5 how to fix and finish plastering mouldings.

The table below shows how the main headings in this chapter cover the learning outcomes for each qualification specification.

Chapter section	Level 1 Diploma in Plastering (6708-13)	Level 2 Diploma in Plastering (6708-23) Units 225 and 226	Level 2 Technical Certificate in Plastering (7908-20) Unit 206
Fibrous plaster mouldings	n/a	Unit 225 Learning outcomes 1 and 2 Unit 226 Learning outcomes 1 and 2	Topics 1.1, 1.2
Select materials, components and equipment	n/a	Unit 225 Learning outcomes 1 and 3 Unit 226 Learning outcome 3	Topics 2.1, 2.2, 3.1, 3.2, 3.3
Plaster reverse moulds	n/a	Unit 225 Learning outcomes 4 and 5	Topics 4.1, 4.2, 4.3
Cast from a reverse mould	n/a	Unit 225 Learning outcome 6	Topics 4.4
Fix and finish plaster mouldings	n/a	Unit 226 Learning outcomes 4, 5, 6, 7 and 8	Topics 5.1, 5.2, 5.3, 5.4

1 FIBROUS PLASTER MOULDINGS

Fibrous plasterwork developed as a modern replacement and a lighter reinforced option for ornate plasterwork that was formed *in situ*. It resembles classic period designs which were created in the Georgian, Regency, Victorian and Edwardian eras.

The process for fibrous plasterwork is to produce reinforced positive plaster casts from a reverse mould.

This type of work is produced and manufactured by a fibrous plasterer in a purpose-made fibrous workshop.

KEY TERM

In situ: when a plaster moulding is run directly to the background, using a positive profile.

▲ Figure 6.1 Different mouldings in position

Benefits of fibrous plasterwork

All plastering work carried out on a plasterer's workbench falls into the category of fibrous plastering. Fibrous plasterwork has benefits that appeal to architects, project managers and plasterers:

- The fibrous items can be manufactured by specialist plasterers in a controlled factory environment (as well as on site).
- Fibrous plasterwork is durable and lightweight.
- The skill and efficiency of the specialist plasterer saves time.
- The project manager can plan the manufacture and delivery of the fibrous plasterwork to meet contractual timescales.
- The quality of finished work can be maintained by following proper fixing procedures.
- It can be manufactured to order beforehand, reducing standing time.
- It can be manufactured to the required lengths.
- It is lightweight and uses less material than solid plasterwork.
- Ornate designs can be manufactured, using various moulding techniques.
- Positive moulding is run on a bench to the shape of the finished product, such as a panel mould or a dado rail.
- Negative moulding is run on a bench as the reverse design, then a positive cast is taken to produce the finished product, such as a cornice.

▲ Figure 6.2 Positive and negative moulding run on a bench

When producing positive mouldings, it is important to understand the design process:

1. Establish the finished pattern.
2. Produce a reverse of the pattern from a running mould **profile**, produced from a drawing or **squeeze**.

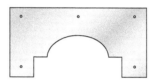

▲ Figure 6.3 A profile template

What is a reverse mould?

Running moulds fall into two basic categories:

- **Positive running moulds** are used to produce a section of the mould on a bench as a **run cast** or *in situ*.
- **Reverse or negative moulds** are run on a bench and a positive plaster cast is taken from the reverse mould.

Reverse moulds are the 'back to front' version of the design shape or pattern you want to produce.

> **KEY TERMS**
>
> **Profile:** the shape and pattern of a mould outline.
> **Squeeze:** a method for reproducing a mould outline.
> **Run cast:** a plaster moulding run on a bench with an upstand to produce a positive profile.

229

Early reverse moulds tended to have plain, simple member designs, but as fibrous plasterers' skills developed over time, undercut mouldings were manufactured using loose piece moulds and flexible compounds.

Insertion reverse moulds were developed, originally using wax and gelatine, to produce ever-more ornamental cornice designs. However, today we use flexible hot or cold compounds and fibreglass to produce detailed ornate designs.

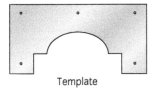
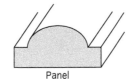

Template   Panel

▲ Figure 6.4 Positive panel mould

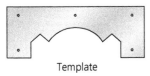
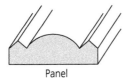

Template   Panel

▲ Figure 6.5 Reverse panel mould

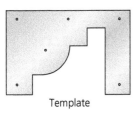
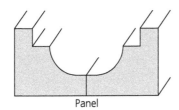

Template   Panel

▲ Figure 6.6 Cornice panel mould

The profile designer who creates the detailed drawing and the plasterer who will manufacture and fix the final item must both have the ability to interpret the drawing and visualise the finished product. The plasterer must also have the skills to accurately transfer geometric mould outlines from a drawing onto a sheet of zinc or aluminium.

When drawings are not available, for example when working for a domestic client, the plasterer will be required to reproduce the profile from an existing cornice. This can be achieved by taking a squeeze.

There are several methods for taking a squeeze:

- taking a plaster squeeze
- using a pin profiler/profile gauge to form the shape of the cornice profile
- cutting into the existing moulding using a fine-toothed saw, inserting a sheet of card and drawing around the profile onto the card
- cutting a small section out of the cornice and transferring the shape directly to a sheet of zinc or aluminium.

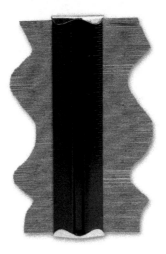

▲ Figure 6.7 Profile gauge

▲ Figure 6.8 Plaster squeeze

Once the fibrous casts have been produced, it is good practice to let them dry to allow excess water to evaporate. Always store the casts flat or upright to prevent distorting and warping. Once dry, they become stronger and are ready to be delivered to site where they can be installed.

Geometrical setting out of moulding outlines

Geometrical outlines for cornice moulding are based on either Greek or Roman architecture. Although many of the shapes are similar, the Greek style is more flowing whereas the Roman style is proportional and based on squares.

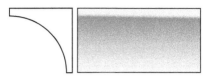

▲ Figure 6.9a Cavetto

▲ Figure 6.9b Ovolo

▲ Figure 6.9c Cyma-Recta

▲ Figure 6.9d Torus

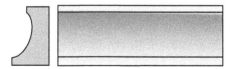

▲ Figure 6.9e Scotia

Information sources used when repairing and making good plaster mouldings

Drawings

Drawings show the cornice design with dimensions of **depth** and **projection**. The main drawings used by a fibrous plasterer will be:

- **detailed drawings**, for the manufacture of the running mould, including the zinc or aluminium profile (usually at 1:10, 1:5 or 1:1 scales)
- assembly drawings for the fixing of the cornice into position on site (usually at 1:20, 1:10 or 1:5 scales).

Schedule

The schedule identifies the room location to ensure each cornice is fitted in the correct room. This is especially important if the site has multiple rooms with different fibrous plasterwork designs to be fitted.

Specification

The specification gives guidance on the required fixing method and the materials to be used for fixing the cornice or other fibrous work, such as a niche. For example, guidance on the required fixing method might include information about the depth and projection of the cornice and instructions, such as:

- Mark out to prepare ceiling and walls to receive fibrous cornice by scoring to improve mechanical key.
- Drill **pilot holes** every 600 mm through timber lath bearers.
- Apply approximately 3 mm of FiberFix adhesive to back leading edge of cornice.
- Firmly push cornice into position and fix with 50 mm drywall screws into plugged or timber background.
- Remove excess FiberFix. Make good and joint as work proceeds.

KEY TERM

Pilot hole: a small, pre-drilled hole bored to help prevent splitting.

▲ Figure 6.10 A niche is a type of fibrous work

On receipt of the contract drawing and specification, it is good practice to review the information to check whether the item on the drawing can be made and that the specification reflects the job to be carried out. Also check that the materials specified comply with the manufacturer's own guidelines.

Any discrepancies should be reported to your line manager if you are employed or, if you are a sub-contractor, to the main contractor or the architect.

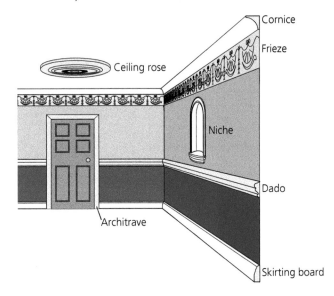

▲ Figure 6.11 Different types of fibrous plasterwork

Figure 6.11 is an artist's impression of a room showing various types of plaster moulding. Although not a working drawing as such, architects often use this type of drawing to illustrate to a client how a room might look once it is completed. You will notice that the picture does not include the measurements or other information a plasterer would need to make and fix fibrous work.

ACTIVITY

Have a look at the decorative plaster frieze in Figure 6.11. What type of mould would you use to produce this decorative fibrous plasterwork?

ACTIVITY

1 Search online to find prices and measurements for the following:
   - plain plaster ceiling rose (find cost and diameter of ceiling rose)
   - plain plaster cornice ogee (find cost, depth and projection of cornice)
   - egg and dart plaster cornice (find cost, depth and projection)
   - decorative plaster frieze (find depth of frieze)
   - solid based plaster niche (find cost, width and length of niche)
   - egg and dart plaster dado (find cost).
2 Sketch or describe the egg and dart design in your own words.

Manufacturer's specifications

The manufacturer's specification will give information such as:
- product/material description and make
- how to apply or install the material or product
- standards of workmanship.

Manufacturer's instructions

The manufacturer's instructions will give information such as:
- technical data, including limitations on use
- how to dispose of waste material in a safe and environmentally friendly manner
- any health and safety considerations while using the material.

Data sheets

Data sheets are produced by the plaster manufacturer and are usually available online. If an employer is planning to use new materials for the first time, they are responsible for informing their workforce how to use them safely. This could be done during a toolbox talk.

Data sheets provide lots of information about the use of the material and usually contain at least 15 important headings:

1 **Product name and company information:** provides a general description of the material and its use, plus the name and address of the manufacturing company.
2 **Hazards identification:** outlines any known hazards when using the material, which can be used when creating the job's risk assessment.
3 **Composition of material:** states the ingredients used in the product.
4 **First aid measures:** gives first aid advice, such as how to deal with skin exposure or inhalation/ingestion of the material.
5 **Firefighting measures:** states whether the material is flammable and how to deal with it in the event of a fire.
6 **Accidental release measures:** gives advice on what to do if the materials spill on the floor.
7 **Handling and storage:** covers how best to handle and store the material.
8 **Exposure control:** states limits of exposure when using the material and engineering controls to minimise the risks when using the material.
9 **Physical and chemical properties:** gives the properties of the material under various conditions.
10 **Stability and reaction:** provides information about the stability of the material in various conditions and how it reacts when mixed with other materials.
11 **Toxicological information:** gives advice on whether the material is poisonous.
12 **Ecological information:** covers information about the effects of the material on wildlife and the natural environment.
13 **Disposal considerations:** recommended methods for safe disposal of the material.
14 **Transport information:** provides information about whether the material requires any special delivery precautions, i.e. whether it is a Department of Transportation (**DOT**) hazardous material.
15 **Regulatory information:** outlines regulations that might apply to that material.

KEY TERM

DOT: stands for Department of Transport; transported materials are given a DOT rating to indicate how hazardous they are.

DOT rating

Material such as sand is low risk, whereas petrol might be considered high risk should an accident occur.

Most lorries have a notice on the back including a DOT number. If the lorry is involved in a traffic accident, this number will help the fire service choose the best option for dealing with a spillage or fire.

Types of backgrounds

There are several different types of substructure background that may be used for *in situ* moulding work, including:

- masonry
- timber laths
- EML
- plaster.

Checklist before starting a repair

Before repairing a damaged moulding, you should consider:

- the extent of damage
- the methods of repair that are going to be used
- if the repair is to a **listed building**, how the repair will fit with the rest of the building
- the location of the moulding to be repaired
- what needs to be done to protect the work area.

KEY TERM

Listed building: a building of particular interest, architecturally or historically, which is considered to be of national importance; details of these buildings are recorded on national lists.

The main methods of repairing damaged plain plaster mouldings are:

- run *in situ*
- run cast
- reverse mould and cast.

The method chosen will be the one that best matches the original design. The repair is made by taking a squeeze then replacing the damaged section and making good.

Calculating quantities

There are many factors to consider when ordering materials for fibrous work, including the width, depth and length of the reverse mould or fibrous cast. Many different materials are used in the process of manufacturing a reverse mould and cornice, such as:

- casting plaster
- retarder, such as glue size or trisodium citrate (sodium citrate)
- tallow
- timber lath
- plywood or timber
- hessian
- fibreglass
- sheet zinc or aluminium
- small tacks
- drywall or wood screws
- shellac
- clay.

Calculating and estimating

Measuring for fibrous plasterwork is carried out using standard measurements which are as follows.

- **Area:** used to measure most plastered surfaces, such as two coat plasterwork, dry lining and finishing plasters. The common unit symbol is m^2 (a metre square).
- **Volume:** used to measure area and thickness of application in three coat work. Plasterers also use volume when gauging loose plastering materials to ensure equal measuring. The common unit symbol is m^3 (a metre cubed).
- **Linear:** with the exception of very deep or complex cornice moulding, the majority of cornice is measured by the linear metre run. The common unit symbol is L m.

INDUSTRY TIP

- Plastering work is measured in area, volume and linear measurements.
- Laths and rolled materials such as hessian are measured by linear measurements.

Sometimes you will need to work out the individual cost of an item that is normally sold in packs, including the VAT cost and any delivery costs. Let's look at an example.

EXAMPLE

If a bundle of 100 laths costs £33.00, how much does each lath cost if you also add VAT at 20% and £15.00 for delivery of the bundle?

Step 1

Work out the cost for the bundle with VAT. In your calculator, enter the bundle cost and multiply it by itself plus 20%. This is written as '1.20'.

$33 \times 1.20 = £39.60$

Step 2

Add the delivery charge to the total cost so far:

$39.60 + 15 = £54.60$

Step 3

Divide the total cost by the total number of items in the bundle:

$54.60 \div 100 = 0.546$

Round this up to the nearest penny and the answer is **55p per lath**.

An easy way to divide by a decimal number without using a calculator is to count the zeros and move them to the left. For example, there are two zeros in 100. Move the decimal point in £54.60 two spaces to the left and you get £00.546. Round this up to £00.55 (55p).

ACTIVITY

Using the internet or trade catalogues, find the price and weight of a bag of fine casting plaster.

2 SELECT MATERIALS, COMPONENTS AND EQUIPMENT

Equipment to produce fibrous moulding work

The majority of fibrous work is carried out in a fibrous plasterer's workshop. Ideally this will have adequate space to store all the materials used to produce moulding work, as well as drying areas for the finished work. If hot melt compounds (HMCs, also known as hot pour compounds) are to be used, adequate extraction must be installed to remove hazardous fumes when melting.

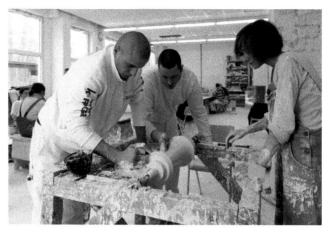

▲ Figure 6.12 A plasterer's workshop

To produce fibrous work, a plasterer needs a strong workbench to support the weight of the materials used.

▲ Figure 6.13 A plasterer's bench

Traditionally a plasterer's bench was made with solid timber legs and an over-boarded top which received 50–75 mm of casting plaster, producing a hard flat surface. Today, most plasterer's workbenches are marine plywood instead of having a plaster top. The most common top size is 1.2 m × 2.4 m, although the size of the top really depends on the type of work carried out. You could easily run a reverse cornice mould from a laminated surface such as a worktop.

The essentials of a plasterer's bench are that:
- it is large enough to produce the item required
- it has running rules on either side of the bench, made from timber or metal
- if the surface is porous, it has been sealed with at least three coats of shellac
- it is sturdy and sits level on the floor.

Some plastering workshops like to use plaster bins for the daily fibrous work, as they keep the plaster dry and promote an organised working environment.

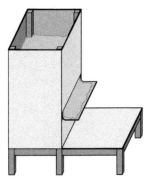

▲ Figure 6.14 Plaster bin

Hot melt compounds

Flexible compounds are used to produce ornate plasterwork. They are made out of polyvinyl chloride (PVC), which is a **thermoplastic** material. This means it has to be heated to between 140–170°C to become a liquid. Always read the manufacturer's instructions for the melting point of the compound.

- Once the compound has reached the required temperature, pour it into a galvanised bucket and leave to stand for about a minute. This allows any air bubbles trapped in the liquid to **dissipate**.
- The hot compound is then poured over a plaster/ clay model surrounded by a temporary fence, or through a metal funnel if the model is surrounded by a plaster case.
- As the compound cools it starts to solidify, forming a flexible rubber compound.
- Leave the hot melt compound to cool overnight.

HMCs are environmentally friendly as the compound can be remelted to produce new flexible rubber moulds.

Grades of HMC

There are three basic grades of HMC and each manufacturer will have its own colour scheme to show the different grades. For example, the manufacturer Vinamold uses red, white and yellow:

- Vinamold Red is a general-purpose PVC and the most flexible compound, suitable for most plaster applications.
- Vinamold White is both flexible and tough, so it is ideal for concrete and polyester resin.
- Vinamold Yellow is the least flexible of the three, so it is used with large moulds and thin sections and is recommended for use with all casting materials.

▲ Figure 6.15 Vinamold

Different grades can be mixed together if required. For example, if you are running short of materials or are just topping up, the different grades are compatible with each other when heated.

How to use HMCs

Follow the steps below when using HMCs.

1. Check that the heating equipment has a current PAT label.
2. Visually check the heating equipment before switching it on. Ideally the heating equipment should be situated under an extractor system, or at least in a restricted, safe and ventilated area.
3. Set the thermostat to the manufacturer's recommended melting point.
4. Estimate the amount of HMC you will need to complete the pour in one attempt.
5. Cut the HMC into small cubes, about 25 mm², as this helps the melting process.
6. Remove any air present in the plaster model by pre-soaking it in water. Do not over-soak, as any visible water on the model surface will be detrimental.
7. Depending on the style and size of the model, form a fence or a case around the model. This is a dangerous operation: you must wear appropriate heat-resistant PPE including arm-length gauntlet gloves.
8. Pour the HMC into a metal bucket and allow it to stand for a few minutes to lose heat. This is a dangerous operation: you must wear appropriate heat-resistant PPE, including arm-length gauntlet gloves. The cooling lets any bubbles in the HMC dissipate. If you pour too quickly, any air bubbles present will appear on the flexible mould.
9. Then, still wearing appropriate heat-resistant PPE, pour the hot HMC liquid in one continuous stream, from the lowest point of the model.

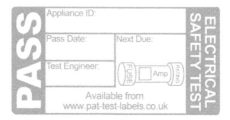

▲ Figure 6.16 PAT label

INDUSTRY TIP

Minimise any possible suction to a bare plaster model by soaking it in water.

Case moulds

- When pouring rubber into a case mould, place a funnel at the base of the model.
- The length of the funnel must be higher than the model otherwise the rubber will not rise to the top. Breather holes at the top of the case will ensure the HMC pushes to the top.
- Cap off the breather holes with clay when the rubber has cooled down.

▲ Figure 6.17 A HMC heater

Flood moulds

- When pouring rubber into a flood mould, remember to make sure that all the porous surfaces have been sealed with a suitable sealer.
- Pour a steady continuous stream of rubber at the lowest point at the base of the model until the rubber is a few millimetres short of the clay, plasterboard or timber fence that surrounds the model.
- Once the rubber is set, carefully remove the fence and the rubber from the model and wash the rubber reverse. Plaster casts can then be taken using one gauge of casting plaster.

Cold pour compounds

Cold pour uses a flexible silicone, two-part moulding compound consisting of a liquid and a **catalyst**, colour-coded to visually aid thorough mixing. It is important to read the manufacturer's instructions before mixing as each silicone compound will have a different mix ratio, depending on the weight of the material.

Working time can vary from 15 minutes up to 90 minutes, depending on the selected silicone. Curing/demould time is at least 10 hours for most silicone compounds.

▲ Figure 6.18 Cold pour rubber compound

Latex rubber is also a cold pour compound. This material does not use a catalyst to activate the set and is brushed over the model to build up as many coats as required. Because of its **thixotropic** properties, it can be applied vertically.

KEY TERMS

Catalyst: commonly used with fibreglass materials or cold pour rubber, the catalyst in liquid form is carefully measured into the bulking liquid. It reacts with the other liquid: in fibreglass it hardens, whereas in cold pour rubber it turns the liquid into a flexible rubber.

Thixotropic: material that remains in a liquid state in its container, but changes into a gel-like state and hardens into position when brushed vertically or horizontally.

Zinc: a non-ferrous metal, zinc is easy to cut and forms its own protective layer called patina, so it does not rust.

Running moulds

Materials

Running moulds are constructed from timber. They use a **zinc** or aluminium template that forms the profile shape of the moulding. You will need the following materials to make a running mould.

▲ Figure 6.19 Cutting zinc

Material	Use
Timber or plywood	To make the running mould.
Zinc or aluminium	To make running mould profiles.
Small tacks or panel pins	To fasten zinc profiles to timber.

Material	Use
Drywall or wood screws	To assemble running moulds.
Shellac	To seal porous plaster.

▲ Figure 6.20 Bench with vice

Tools

You will need a workbench with a vice to hold the timber and zinc steady while the materials are cut, shaped and assembled. As well as this, you will need the tools in the following table for the running mould making process.

Tool	Use
Tin snips	Basic or avionic tin snips are used to shape the zinc or aluminium into a mould profile.
Files	Files are required to form both the timber and the zinc. Files come in a variety of styles and sizes. You will need at least a flat and a half-round fine metal file for shaping the zinc or aluminium and similar wood rasps for forming the timber.
Wood saw	A sharp wood saw is used to cut the timber or plywood **horse**, **stock** and **brace** to size.
Coping saw	Once the profile has been cut to shape, the outline of the mould is placed on the timber/plywood stock and the outline is extended by approximately 3–5 mm. The coping saw is then used to shape the timber.

Tool	Use
Tenon saw 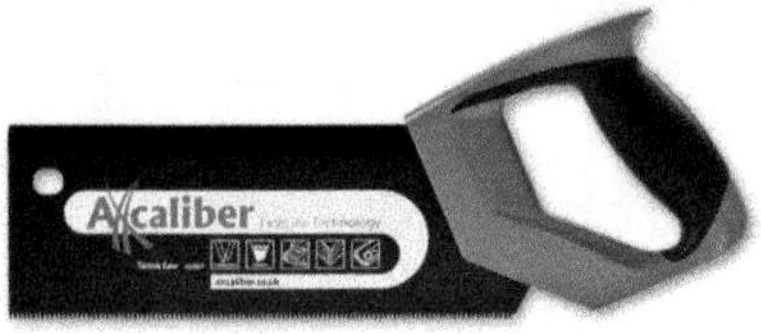	This fine-toothed saw is useful for cutting rebates into timber.
Pein hammer 	This small lightweight hammer is useful for securing the zinc profile onto the timber stock with nails. Also known as a pin hammer.
Drill/driver 	A small lightweight drill/driver is essential to drill pilot holes and secure the running mould, together with suitable screws.
Try square 	Also known as a carpenter's square, it is useful for making sure the horse is square with the stock and if you decide to cut a housing joint into the horse instead of a butt joint (see page 247).
Sandpaper/wet and dry emery paper	Sandpaper is used to smooth down rough edges of timber. Wet and dry emery paper can be used to smooth out high spots on zinc or aluminium profiles.
Busks 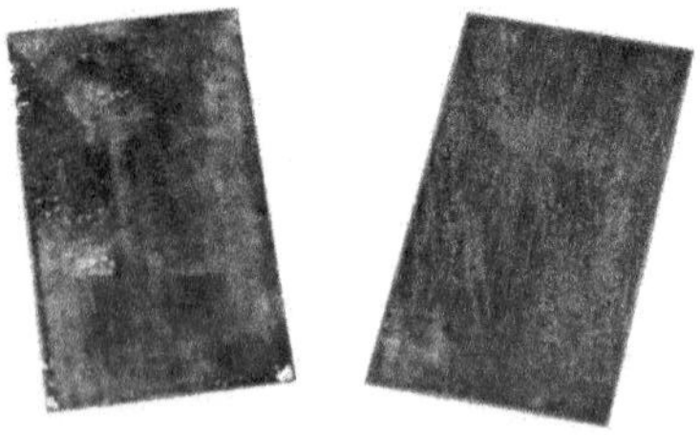	Sometimes known as 'drags', busks are small pieces of thin flexible metal available in three shapes: square, kidney and rectangle. Their main use is for cleaning mouldings and for making good joints.

Tool	Use
Small tools 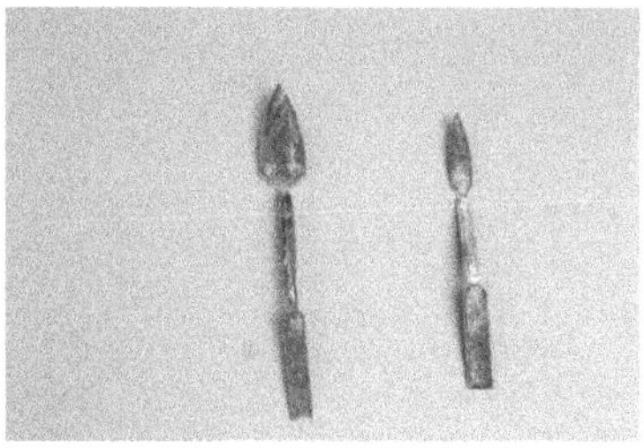	An essential part of the plasterer's tool kit, available in various shapes and sizes, the most popular are trowel and square and leaf and square. Useful for mixing small quantities of plaster, for making good damaged plasterwork and mitre and for working behind pipes.
Gauging trowel 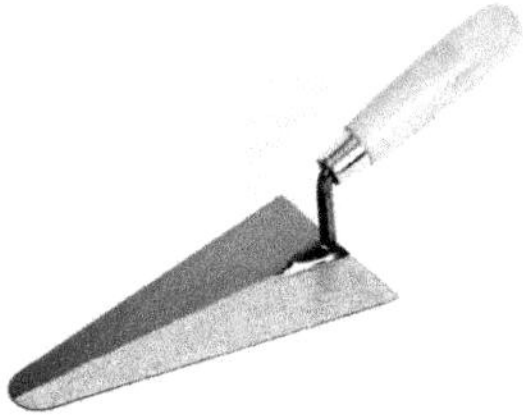	Also known as a 'bull nosed trowel', gauging trowels are used to mix small quantities of plaster and feed plaster into awkward areas.
Splash brushes 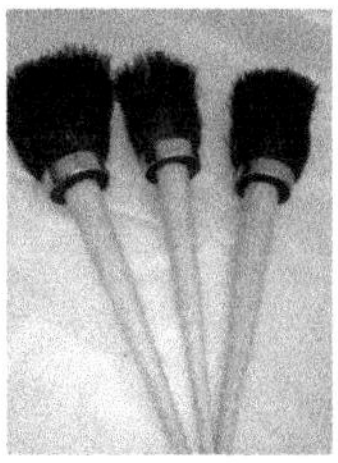	Traditional splash brushes have coarse bristles and are circular, unlike standard paint brushes which have various types of bristles and are oblong in shape. A splash brush is used to brush plaster into **enrichments** in reverse moulding and then to brush and splash plaster when casting.
Shellac brush 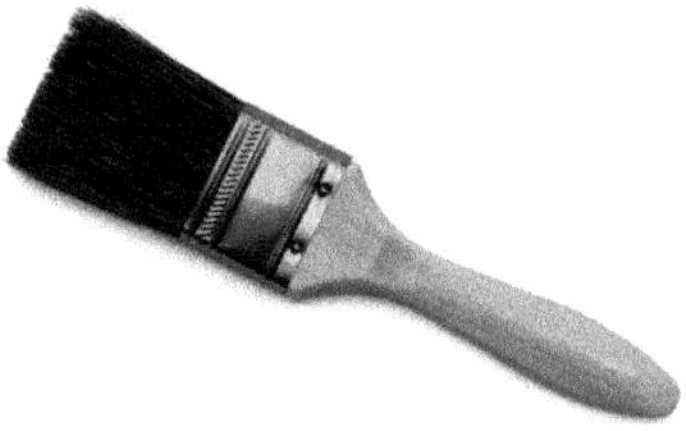	Any good quality paint brush can be used to apply shellac. Any size can be used but 50 mm is perhaps the most common. The most important issue is not to leave the shellac brush exposed to the air for too long or the brush will be ruined. Always leave the brush in water after use or in a shallow container with shellac in it.
Mixing bowls 	Used for mixing small quantities of plaster. As bowls are designed to be flexible (unlike plastering buckets), casting plaster can be left to set hard in them because it will just pop out.
Scraper 	Used to clean plaster from floors. The blade of a scraper can also be used as a joint rule/busk for cleaning plaster slabs and mouldings.

3 PLASTER REVERSE MOULDS

Materials

The plaster used in fibrous work is commonly known as casting plaster, although its proper name is hemihydrate plaster. It is made from gypsum, a naturally occurring mineral rock with the chemical calcium sulphate.

- The rock is extracted from underground mines or open cast quarries and the crushed gypsum is fed into containers called kettles.
- It is then heated for about two hours at a temperature of 170°C, until three-quarters of the waters of crystallisation have been driven off.
- The resulting material after grinding is the hemihydrate plaster, also known as 'plaster of Paris'.

When plaster is added to water, the reverse chemical reaction takes place and the waters of crystallisation re-form to convert the plaster back into gypsum. Chemically, the plaster will accept the exact amount of water that was driven out during the calcining or heating process; any excess water will be left to dry out later.

You will notice that the plaster becomes warm as it sets; this is part of the chemical reaction. Do not attempt to remix gypsum plaster as it sets because the crystals will not lock together, resulting in a weak final set.

Manufacturers of casting plaster will formulate products to suit specific applications. By doing so, they create products with the desired properties, such as set time, fluidity and strength.

▲ Figure 6.21 Casting Plaster

Types of casting plaster

In Europe, a major manufacturer of casting plaster is Saint-Gobain Formula. The characteristics of casting plaster depend on its water/plaster (W/P) ratio, hardness and compressive strength. As a rule of thumb, the more water required to mix the plaster, the more fluid the mix will be, but the strength will be reduced and the setting time extended. Conversely, with less water the setting time is shorter, but the hardness and compressive strength are greater.

- Fine casting plaster 100/70 = 100 parts by weight of plaster to 70 parts by weight of water – this is a weaker plaster.
- Crystacal Alpha K 100/21 = 100 parts by weight of plaster to 21 parts by weight of water – this is a harder plaster.

The method of manufacture also determines the ultimate strength and water ratio of the plaster.

- If the plaster is manufactured in industrial kettles under atmospheric pressure, it is classed as a beta plaster. This plaster will be cheaper but weaker.
- Alpha plaster is heated to the same temperature as beta plaster (200–392°C), but is made in an autoclave oven under steam pressure, which gives the plaster its hardness.

Often, two different mixes of plaster are used for casting. The first mix is known as the firstings and the second mix is known as the secondings or seconds.

Trade names

- *Prestia Classic Plaster (100/70)*: a consistent general-purpose casting plaster, ideal for fibrous plasterwork and sculptural work.
- *Prestia Normal Plus Plaster (100/66)*: similar to Prestia Classic but with a slower setting time; useful when a longer working time is required.
- *Prestia Casting Plaster (100/68)*: ideal for general casting, cast moulds and curving.
- *Formula Fine Casting Plaster (100/70)*: probably the most common general-purpose plaster, it is a versatile and economical beta plaster and is used in numerous industries.
- *Formula Fine Casting Plus (100/70)*: used to make decorative fibrous plaster pieces such as coving, cornices and columns.
- *Formula Keramicast (100/56)*: a hard plaster used for bulkheads, columns, lighting profiles and dado rails.
- *Prestia Traditional Plaster (100/70)*: a slow-setting general-purpose beta plaster.
- *Prestia Creation (100/50)*: a good quality hard plaster useful for most casting and modelling applications.
- *Formula Herculite No. 2 (100/42)*: a hard plaster used for high strength and surface durability.
- *Formula Herculite Fibrefix* (100/55): an adhesive for fixing all plasterwork as well as for filling and stopping. Used on-site where strong adhesion and longer working time is necessary.

- *Formula Crystacast (100/28)*: an extremely hard plaster, used where exceptional detail is required.
- *Formula Crystacal R (100/35)*: an extra hard plaster used for its high strength and hardness.
- *Formula Crystacal Alpha K (100/21)*: ultra high-strength casting plaster that can be used for industrial modelling.

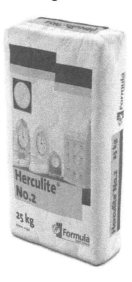

▲ Figure 6.22 Formula Herculite No. 2

ACTIVITY

Plaster companies sometimes merge with other companies or are taken over, but except for some new branding, the products largely stay the same.

In pairs, research the latest information on three of the products in this list.

IMPROVE YOUR MATHS

Plaster to water ratio by weight is 1.43 kg/litre. This means that if you weigh 1.43 kg of plaster, you will need 1 litre of cold water to mix the plaster for it to reach full strength. If you use too much water, the plaster will be weaker. Sometimes the ratio might be shown with decimals, worked out by dividing 1 by 1.43:

$$1 \div 1.43 = 0.70$$

This is then referred to as 100/70.

Materials used for reinforcement

Casting plaster can be brittle. To improve the overall strength of fibrous work, timber, hessian and fibreglass strands can be used.

Reinforcement material	Use
Hessian	Hessian comes from the jute plant. It is woven into coarse fabric that is available in many sizes, usually in 200 m rolls and in widths of 75 mm, 100 mm, 150 mm, 300 mm and 450 mm. It is still the most common material used in traditional fibrous work. Hessian can be wrapped around damp timber laths or wire to improve their **flexural strength**. You can form hessian ropes and wads by dipping hessian into wet casting plaster. Remove the excess plaster by running your fingers loosely over the hessian. Then place the hessian in position from fibrous slabs to suspended fibrous slab ceilings.
Sisal	Sisal comes from the sisal plant. Its main use is to manufacture twine and for making rope. Like hessian, sisal is a popular choice as fibrous reinforcement material.
Fibreglass	Chopped fibreglass is a more modern material used to reinforce fibrous plaster. It is useful to tuck into awkward places, where hessian may be difficult to drape. Chopped fibreglass can be bought in lengths from 3 mm to 20 mm; 15 mm is the most popular size. Fibreglass can also be added to backing and finishing plasters and to floor screed mixes to help prevent cracking. Chopped fibreglass and matting are most often used in acrylic resin-based plasters that can be used externally, such as Jesmonite or Fibrocem.
Timber lath	Softwood timber lath can be used along with hessian to reinforce fibrous casts, providing a strong, rigid skeleton and preventing casts or mouldwork from snapping. Timber laths are generally 3 mm and 5 mm thick (or deep). Widths may be 10 mm, 13 mm or 22 mm and lengths are either 2.4 m or 3 m. The most popular size for fibrous cornice work is 3 mm deep by 22 mm wide by 2.4 m long. Laths require soaking overnight to prevent the cast from cracking when it dries out.

IMPROVE YOUR MATHS

1 Measure the perimeter of a room in your home.
2 Calculate the number of timber laths required to produce enough 3 m lengths of fibrous cornice, if two lengths of laths are required for each 3 m length of cornice.

Sealants and release agents

As well as casting plaster and reinforcements, you might need to use some of the following sealants and release agents.

Shellac

Shellac is a traditional material used to seal plaster. It is a resin secreted by the female lac bug on trees in the forests of India and Thailand. It is processed and sold as dry flakes.

When the plaster reverse mould has been run on the bench, before any casting can take place, the plaster needs to dry and be sealed, because the plaster surface is porous. Shellac flakes are dissolved in methylated spirits, usually overnight. The shellac is then brushed onto the plaster surface and left to dry.

Second and third coats of shellac are applied after the previous coat has dried. More coats of shellac can be applied to plaster reverse casts, depending on the **viscosity** of the shellac and the **porosity** of the reverse cast.

You can also use shellac to seal your workbench.

▲ Figure 6.23 Shellac

INDUSTRY TIP

It is better to have the shellac weaker rather than stronger, because it needs to be absorbed into the porous plaster reverse mould.

ACTIVITY

A 1-litre container is half-filled with shellac flakes. How many millilitres of methylated spirit will be required to give a 50:50 ratio?

Tallow

Tallow is solid animal fat from beef or mutton. The solid fat is warmed up gently by heating it in a pan. As soon as the fat has softened and become runny, it is removed from the heat and paraffin is added at a ratio of approximately 50:50. At this stage in the process, it is known as 'plasterer's grease'.

Plasterer's grease is used as a **release agent** to help prevent plaster from sticking to the workbench or the plaster cast sticking to the reverse mould. During cold weather, the plasterer's grease might start to solidify and in summer it might become too runny. To solve this problem, adjust the amount of paraffin added to the softened tallow, or store the grease in a warmer or cooler area of the workshop, as necessary.

Other materials can be used as an alternative to plasterer's grease, such as lard, vegetable oil and linseed oil. For exceptionally fine-detailed cornice works, oil-based materials are a better choice as a release agent because of their thinner consistency.

KEY TERMS

Viscosity: how thick or runny a liquid is. A viscous liquid is thick and sticky and does not flow easily.

Porosity: how porous a material is; a porous material has many tiny holes or 'pores' in it and will easily absorb air or liquid.

Release agent: a substance applied to the surface of a mould to make it easier to remove after the plaster has set.

▲ Figure 6.24 Tallow/plasterer's grease

French chalk

This is a fine powder, similar to talcum powder. It is used when the reverse is greased to find any spots that have not been greased.

RL247 wax

This type of wax is a brush-on liquid wax release agent, also available in an aerosol canister.

▲ Figure 6.25 RL247 wax

Retarders

Retarder can be added to casting plaster to slow its setting time.

Glue size plaster retarder

This retarder is made from animal skins and hooves and is traditionally used within the ornate plasterwork industry. When added to gauging water, it slows down the set of the plaster, allowing you more working time.

To make a batch of this retarder (called 'size' in the trade):

- add 1 kg glue size crystals to a gallon bucket of very hot water
- stir the crystals into the hot water to dissolve; do not allow the water to cool
- while the water is still warm, add approximately 25 g of hydrated lime and stir until fully dispersed
- allow the mixture to cool.

Keatin is very similar to glue size and is also used as a retarder.

Trisodium citrate

This retarder is very easy to make:

- Add 1 kg to a gallon bucket of very hot water.
- Mix until fully dissolved and then allow to cool down.

Components of a running mould

Running moulds can be constructed in a variety of shapes and sizes. The profile design determines the moulding member features of both positive and negative profiles.

Running moulds must be robust, strong and constructed from the correct materials in line with specifications, to ensure they are fit for purpose.

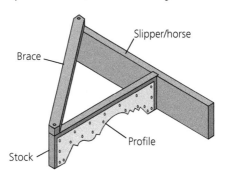

▲ Figure 6.26 Parts of a running mould

The basic components of a running mould are as follows:

- Horse or slipper: two different regional words to describe the same part. The stock is fixed to the horse/slipper with nails or screws. The horse then runs down the bench against the bench's running rule.
- **Stock:** the stock is fixed at 90° to the horse/slipper with a butt or housing joint (see Figures 6.28 and 6.29). The stock is approximately 3 mm larger than the profile. This helps prevent a build-up of plaster on the back of the profile when running the plaster mould.
- Brace: timber that is fixed to the horse/slipper and the stock at an angle of 45°, using nails or screws to brace and strengthen.
- Profile: sometimes known as a template, this is made from zinc or aluminium to the shape of the mould (positive or negative) and is fixed to the stock using small tacks or screws.

▲ Figure 6.27 Housing joint

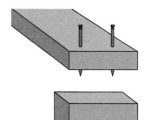

▲ Figure 6.28 Butt joint

Types of running mould

This list shows some of the types of running mould and their uses.

- Plain panel/dado mould: a small positive mould run on a bench.
- Plain reverse mould: a negative cornice mould.
- Loose piece reverse mould: a negative cornice mould with an undercut profile.
- Loose piece insertion reverse mould: a negative cornice mould with ornate detail.
- Extended stock running mould: to run circular moulding.
- Gig stick running mould: to run circular moulding with a larger radius.
- *In situ* running mould: to run a mould direct to the wall or ceiling, *in situ*.

The following step-by-step instructions show you how to construct a running mould. The photographs on page 248 show a positive panel mould being constructed, but the principle is the same for making either a positive or negative mould.

STEP 1 Design the basic shape of the profile on graph paper and stick it onto a piece of aluminium or zinc.

STEP 2 Cut out the shape of the profile with tin snips to within 2 mm of the profile line.

STEP 3 Using a smooth metal file, shape the zinc to match the paper template.

STEP 4 Remove any **burrs** with wet and dry emery paper.

STEP 5 Cut your stock. This should be wider and longer than the profile metal (5 mm bigger in both directions) to fit the profile on its surface. This allows for any swelling that might occur when in contact with wet plaster. Fix your aluminium/zinc profile to the stock using tacks.

STEP 6 The stock and profile can now be fixed to the slipper. For strength, the stock can be notched then glued and screwed into position. Make sure the stock and slipper sit flat on the bench before fixing them together. For stability when running on the bench, it is normal for the slipper to be longer than the stock.

STEP 7 Braces can be fixed to support the stock and slipper. Use a square to make sure the stock is set at a right angle to the slipper before fixing.

STEP 8 The running mould has been constructed but you might want to seal the timber with shellac before it comes into contact with wet plaster. This will increase the running mould's life span and prevent distortion of the timber.

KEY TERM

Burrs: rough edges on the profile left after filing.

Undercut designs

Some cornices may incorporate an undercut design. This design feature makes it difficult to release the fibrous cast from the reverse.

▲ Figure 6.29 Undercut cornice

Scotia

▲ Figure 6.30 A scotia member

Ornate undercut pattern

▲ Figure 6.31 Ornate pattern design

A model or mould with overhanging ornate design is known as an enrichment. When casting a cornice mould with an ornate design, the mould will be impossible to take off: the undercut pattern of the enrichment on the reverse mould will prevent the cast being taken off when set.

In this situation, the **inserted** enrichment on the reverse mould is produced with a flexible compound material:

- Form a plane rebated section in profile of the running mould.
- Bed the enrichments in the rebated section of the positive mould and leave to dry.
- Carry out **fencing** and sealing.
- Pour cold or hot rubber compound over the positive mould to produce a flexible reverse mould.

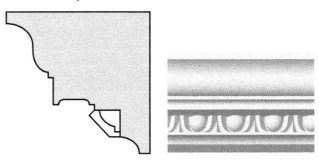

▲ Figure 6.32 Egg and dart decorative plasterwork

KEY TERMS

Insertion: incorporating an enrichment pattern in to a moulding to enhance its design.

Fencing: ensuring that a model such as an enrichment is surrounded, to eliminate leakage when flexible compounds are poured over the enrichment model to form a reverse.

ACTIVITY

1 Search the internet for different moulding designs.
2 Make a list of the different enrichment patterns found on ornate moulding work.

Using a backboard to produce a run cast moulding

Although not often used today due to the amount of labour involved with the process, fibrous plasterers can obtain the same design by carrying out a procedure known as run cast process.

A backboard is formed by securing a length of timber vertically on the plasterer's workbench. It can be used to run a cornice when only small amounts are required for matching existing moulding work or running short breaks that might be difficult to run *in situ*.

Procedure for running a loose piece mould

To run a loose piece mould:

- Prepare the bench in the usual way.
- Run the reverse mould with the detachable loose piece profile in position.
- Apply two to three coats of shellac to the section that the loose piece ran over and grease this lightly to prevent plaster sticking to the loose piece channel.
- Run the reverse mould again, filling the loose piece channel. It is wise to strengthen this section with lath, hessian or fibres.
- Seal the whole of the completed reverse mould with shellac three times.

When a cast is taken from this reverse mould, the loose piece will separate from the mould, allowing the cast to be removed without any problem.

Waste moulds

Waste moulds are used to produce a one-off moulding from a clay model where the sculptor wishes to retain exclusivity.

- Prepare the clay model with a thin coat of a release agent such as linseed oil or Mac Wax spray.
- Mix casting plaster with a coloured tint, such as vegetable dye used in cake-making.
- Lightly splash the mixture onto the clay model to a thickness of approximately 5 mm. Make sure that no air is trapped between the clay model and the tinted plaster and that the model is completely covered with the tinted plaster.
- As the plaster starts to set, brush a weak mixture of clay water over the tinted plaster, just enough to discolour the surface.
- Mix and apply ordinary casting plaster over the clay wash to an approximate thickness of 12 mm, depending upon the size of the model.
- Once set, turn over the mould and carefully pick out the clay original.
- Rinse the inside of the mould with clean water; apply a mixture of soft soap and water to the mould, making sure all of the crevices have a coating of soft soap.

- Once coated, pour a creamy gauge of casting plaster to the top of the mould and gently shake to remove any trapped air.
- The plaster cast must be fully set before you attempt to remove the outer case. To remove the outer case, use sharp chisels and cutting tools to cautiously cut and chip away until the tinted plaster is reached. This indicates that the cast has nearly been reached, so gradually remove the tinted plaster with blunt tools to avoid potential damage to the cast.

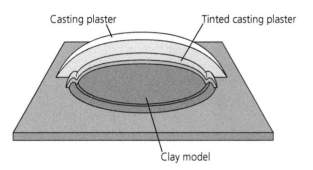

▲ Figure 6.33 Waste mould

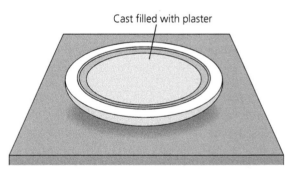

▲ Figure 6.34 Mould turned over with clay model removed and casting plaster inserted

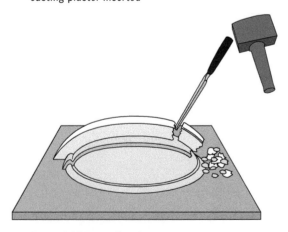

▲ Figure 6.35 Removing the case

Large, more intricate waste moulds are best made in several sections to ease the removal of the clay and then the waste plaster.

Insertion moulds

Insertion moulds can be manufactured in the following two ways.

Method 1

- Run a cornice *in situ* with a channel in the profile.
- Cast enrichments separately from a rubber mould. Bed them into the channel with casting plaster or a suitable plaster adhesive and joint them as required.

▲ Figure 6.36 Run cornice with sinking channel

Method 2

The more popular method for producing a decorative cornice with enrichment moulding is to run a reverse mould with channels and fix decorative moulding in place, then take a rubber reverse moulding from which plaster casts can be taken.

- A modeller will create a clay model of a decorative section of the plasterwork on a sheet of plywood.
- The next stage is to make a clay fence around the model, which should extend to at least 10–20 mm from all parts of the model including its highest part: this is to allow for the pouring of the hot or cold pour compound. The thickness of the clay fence depends on the size of the model: the fence must be robust enough to be free standing and to hold the pressure of the poured compound. Other materials, such as timber or plasterboard, can be used to form the fence around the model as long as they are sealed with shellac.
- In a clay model, make a small section and take a rubber reverse. Make as many plaster casts as necessary to cover a sufficient length of a panel

mould from the small rubber reverse. Join these together to provide a seamless plaster model; again, build a fence around the plaster model ready to receive hot or cold pour compound. These small castings are called flood moulds because the plasterer just 'floods' the rubber reverse with plaster.

INDUSTRY TIP

When casting from reverse moulds, it is important to remove trapped air by vibrating the bench surface during the casting process. Otherwise, there will be imperfections and air holes on the mouldings.

▲ Figure 6.37 Reverse mould with enrichment

Running a reverse mould

Preparation

INDUSTRY TIP

Prepare all resources and material beforehand and keep your working area and running mould clean as you work to prevent unnecessary build-up of plaster.

- Scrape and sweep the bench with a brush to remove any loose materials.
- Check that the running rules are firmly in position and free from any sticking plaster, as this would hinder the smooth movement of the running mould along the bench.

- Repair any indents on the bench surface and apply a few coats of shellac to seal the surface of the bench.
- Check the robustness and strength of the running mould and the sharpness of the profile to prevent any unwanted drag lines that would spoil the finished work.
- Get two buckets of clean cold water, a brush to clean the running mould and at least three clean flexible mixing bowls. Apply some barrier cream or wear a pair of disposable latex-style gloves to protect your hands.
- It is important to mix enough plaster to run the full length of mould, with no major misses along the length and members of the moulding. If this is not done, there will be uneven expansion along the length of the mould during the later stages of the run, when running down to a finish.

▲ Figure 6.38 Cleaning the bench with a scraper

INDUSTRY TIPS

- When bagged materials are delivered, always check the use-by date on the side of the bag. It is not uncommon for suppliers to deliver materials that are out of date or nearly out of date.
- When you store bagged materials, remember to rotate the stock so they are used on a 'first in, first out' basis.

A reverse mould requires a **core** to help reduce expansion of freshly mixed plaster during running by using less plaster. This is a good opportunity to recycle offcuts.

KEY TERM

Core: old moulding or plasterboard incorporated in the reverse moulding; used to reduce expansion because it reduces the amount of plaster required.

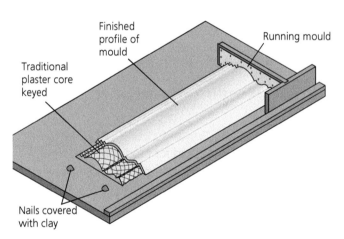

▲ Figure 6.39 A reverse mould with a plaster core

INDUSTRY TIPS

- A running mould can be made either left- or right-handed, but the principle of using the running mould is the same.
- Do not be tempted to add the plaster to the water and mix it straight away, as your mix will be lumpy and inconsistent.

Running the mould

STEP 1 Position the core on the bench.

STEP 2 Fix the core in place.

STEP 3 Apply plasterer's grease to the bench. This will allow the running mould to move freely and act as a release agent.

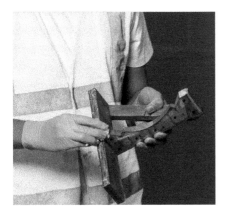

STEP 4 Lightly grease the running mould before using it, to help prevent the plaster from sticking.

STEP 5 Add some clean cold water to a mixing bowl and then sprinkle through your fingers small quantities of casting plaster until the plaster sits just below the water line. Let the plaster soak for a minute or two. Then gently mix the plaster through your fingers to produce a smooth lump-free creamy consistency.

STEP 6 Pour the creamy plaster mix just longer than the required length of the finished mould. This is to allow for wastage.

STEP 7 Place firm downward pressure on the horse against the running rule and the nib of the stock. This ensures the running mould has full contact with the bench and helps prevent plaster building up beneath the running mould.

STEP 8 The zinc or aluminium profile should always face the direction you are working. With one hand on the nib and the other on the brace, walk while pulling the running mould. As you are walking, feed the surplus soft plaster from in front of the profile back into any low spots.

STEP 9 As the running mould is gradually built up, mix smaller wetter quantities of plaster to finalise the running.

STEP 10 Wash the mould. Repeat Steps 8–10 several times. Once you are satisfied that the running mould is complete, leave the mould for 20 minutes or so until the plaster has set. During this time, take the opportunity to clean up and get ready to run the next cast.

STEP 11 Gently run the blade of a small tool to both edges of the reverse mould. This will release the mould from the bench.

STEP 12 Brush the sides of the mould.

STEP 13 When the reverse mould is complete, trim off the rough ends.

STEP 14 Apply three coats of shellac to seal the porous plaster surface. Once sealed, the reverse mould can be used many times to produce fibrous cornices.

Gently tilt the panel mould on its edge to move it to a suitable storage area.

To speed up the running process, a false profile can be built up from plaster, or a zinc plate can be fixed temporarily to the stock profile. This will reduce the amount of casting plaster used and also reduce mass swelling when producing the reverse moulding.

The false profile should extend the original profile by approximately 6 mm and can be made from casting plaster or any sheet material such as zinc, plywood or hardboard. In plastering terms we call this false profile a **muffle**.

KEY TERM

Muffle: a temporary plaster mix applied to extend the profile by 5–6 mm; thin ply or a zinc sheet can also be used.

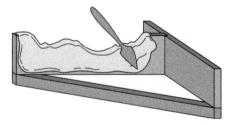

▲ Figure 6.40 Use a small tool to build up the muffle gradually (view from inside the running mould)

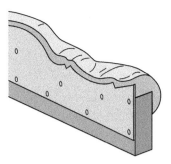

▲ Figure 6.41 Scribe the set plaster to the shape of the profile

Running a plaster positive panel mould

The method used to produce a bench panel mould is similar to that for making a plaster reverse mould – but a panel mould is smaller and does not require a core.

- Drive two headless nails into the bench, approximately 500 mm apart, leaving about 10 mm of each nail projecting out of the bench at the deepest section of the mould.
- Place a small cone of clay over the nail heads. This prevents the expanding plaster moulding from moving when being run and stops the plaster sticking to the nail head.
- Depending on the size of the running mould, you can incorporate reinforcement by using hessian scrim.

To run a panel mould, refer back to pages 247–248 as the step-by-step instructions are the same as for running a reverse mould.

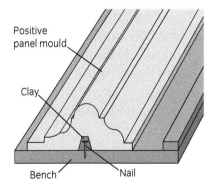

▲ Figure 6.42 Bench mould in place

▲ Figure 6.43 Running a panel mould

Checklist for running a panel mould

- Prepare the bench and running mould: check and repair any defects.
- Grease the bench and running mould.
- Secure headless nails every 500 mm and cover with clay.
- Select clean buckets, bowls, water brushes and water.
- Select a small joint rule.
- Mix plaster to a creamy consistency and pour just beyond the length of the finished mould.
- Build up the running mould shape with creamy plaster mix as quickly as possible.
- Finalise the running mould shape with smaller, wetter quantities.
- Clean up and leave the plaster mould to set.
- Carefully remove the plaster mould from the workbench.
- Place the panel moulds flat in a well ventilated area.

HEALTH AND SAFETY
- Casting plaster is considered a safe material to use. However, you must take care not to inhale its dust: mix it in a well-ventilated area and wear a dust mask. Wear barrier cream or nitrile gloves to help prevent absorption into the skin and also wear safety glasses.
- Never cast body parts on raw skin, especially your hands. Skin burns can occur at 45°C and the cast can reach 60°C when setting.
- Dispose of casting plaster in a separate skip for recycling. Tip dirty plaster water down a filtered trap, but scrape the plaster slops at the bottom of a bucket into the plaster skip.

▲ Figure 6.44 Nitrile gloves

IMPROVE YOUR MATHS

This activity can support numeracy and literacy skills.

Imagine you are working for a fibrous plasterwork plasterer. A new contract has been awarded to your company to make 840 lengths of cornice for a hotel chain. On average, you can make three lengths of cornice an hour and you have a 35-hour working week.

1 How many weeks will it take you to make 840 lengths of cornice for this contract?

2 If the contract value is £14,700.00, what is the unit cost for a length of cornice?

CASE STUDY

Kwame has just left school and is joining a fibrous plastering company as an apprentice.

Kwame will work alongside Sam, who is the supervisor and an experienced **fibre hand** with over 20 years'

experience. After familiarising himself with health and safety during his induction, Kwame's role is to work alongside Sam, producing fibrous casts which incorporate an enrichment for a local client.

As Sam cut out the sink profile for the reverse mould, he explained to Kwame how important it was for the mould to be cut to match the drawing design and specification accurately. Kwame observed and later assisted by sanding the sink profile to remove burr marks which were pointed out during inspection. Sam constructed the running mould and Kwame prepared the bench and resources that he had become familiar with during his first week at work.

As Sam produced the reverse mould, he explained the running process to Kwame. Kwame was allowed to run the mould along the bench at the start of the running process and to form the shape of the reverse mould. Sam supervised and instructed Kwame on his techniques. As the moulding process came to the final stage, Sam complimented Kwame's technique and told him to clean all resources and set up for the next stage.

Sam explained to Kwame that it was his role to manufacture the enrichment mouldings, while it would be Sam's job to prepare the positive mould and insert the enrichments.

Kwame poured plaster into rubber enrichment reverse moulds and Sam inserted them into the mould. However, while inserting the enrichments Sam noticed there were a lot of imperfections and minor air holes on some of the mouldings Kwame had produced. Sam observed Kwame and noticed he was not vibrating the table after pouring the plaster into the reverse. Sam explained to Kwame why it was important to vibrate the table, to ensure the plaster flows into every detail of the reverse, eliminating trapped air and creating a sharp design.

Sam explained to Kwame that this was a common mistake made by an apprentice, but he should not lose heart as he was making excellent progress as a trainee fibre hand.

KEY TERM

Fibre hand: a person who manufactures and installs fibrous plaster components.

Kwame and Sam went on to prepare and complete the positive mould and produced an excellent reverse mould containing an enrichment modillion block.

▲ Figure 6.45 Modillion block

4 CAST FROM A REVERSE MOULD

The next stage is to make a fibrous cast from the reverse mould and then transport the finished items to the client for fixing.

▲ Figure 6.46 Casting a ceiling rose

Suitability of materials for fixing the moulding

It is essential that all the materials used are compatible with each other, including the background that the fibrous work is being fixed to. Remember to follow manufacturers' instructions when preparing and installing fibrous plasterwork.

Mechanical fixings must be rust-free to avoid staining the finished work. All types of drywall screws are suitable, but zinc-plated screws offer better all-round performance. In the same way, if any fibrous components are used to fix fibrous work with the wire and wad method (see page 270), the wire must be strong enough to support the fibrous component and be galvanised to prevent rusting.

FiberFix is a material commonly used for direct bonding fibrous work to solid backgrounds. Coving adhesives can be bought in powder or pre-mixed form. However, they have very different properties with respect to setting times, fineness of the materials and adhesive strength. With some powder adhesives, there is no need to prepare the background with PVA.

IMPROVE YOUR MATHS

A room in a Victorian house measures 3.85 m × 4.25 m.

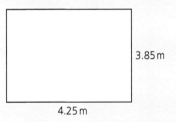

3.85 m

4.25 m

1 How many lengths of plain-faced coving are required if the lengths are manufactured 3 m long? Remember to round up to the nearest whole number.

2 The coving costs £8.00 per length and 10 kg of coving adhesive costs £12.30. The cost of fixing the coving is £3.50 per linear metre. Including all jointing, how much would the job cost, *excluding* profit and VAT?

The specification

You should refer to the specification to check whether the fibrous work is to be mechanically fixed or fixed using adhesive. It could be a combination of both fixing methods. When fixing cornice work, you should also check projection and depths and the position of dado or panel mouldings.

The manufacturer's instructions and the specification will also cover how to prepare the background to receive the fibrous work.

For installation instructions, the schedule states the fixing locations and positions of the fibrous work.

▲ Figure 6.47 Measuring projection

INDUSTRY TIP

Order the moulding work to site in line with the programme of work, as this will reduce standing time, material damage and storage problems resulting in wastage.

You should also consider protecting the finished work once it is fixed, especially exposed items of work such as dado or panel mouldings. Any work damaged on site without being signed off could lead to a contractual dispute over payment.

Part of the challenge of working on site is planning to receive materials from stockists. These materials can include fibrous work for fixing, if these items are not being manufactured in house, and fixing materials such as adhesive, wire and fixing screws.

- On larger sites, there is usually a designated person to receive all inward goods and materials.
- On smaller sites, the delivery driver will probably deal directly with the subcontractor.

The principles for receiving goods are the same for a large or small site:

- Check the delivery note against the official order. Check especially that the quantity and grade delivered match the quantity and grade ordered.
- Check the goods for any signs of damage and, if there are no problems, sign the delivery note to acknowledge receipt.
- Any discrepancies found with the delivered goods, such as damaged goods or shortages, should be reported to your line manager so they can decide whether to take part-delivery or return the full consignment. Remember that once you sign for the goods, your company is responsible for payment under the terms and conditions of the supplier's contract.
- Store materials in a secure location that is appropriate for the type of goods received:
  - High-value goods should be stored in a lockable container; it is good practice to store less valuable but equally desirable items such as screws in a lockable container, too, as these items have a tendency to slip into overall pockets.
  - Perishable goods, that is bagged materials with a use-by date, must be stored in rotation, so that older materials are used first ('first in, first out').
  - Other materials, such as timber, fibrous work and sheet materials, are usually stored off the ground, horizontally or vertically, in a dry and well-ventilated secure storage compound.

The casting process

Select materials, components and equipment

The following materials are used to cast fibrous plasterwork:

- casting plaster
- release agents
- sealing agents/shellac
- reinforcement.

▲ Figure 6.48 Mixing casting plaster in a bowl

IMPROVE YOUR MATHS

1 What is the number given as pi?
2 Draw the symbol for pi.
3 A ceiling rose has a circumference of 1005 mm. What is the diameter of the ceiling rose?

ACTIVITY

Search online for 'hemihydrate plaster' to find the chemical name for casting plaster.

Tools and equipment

The tools used to fix fibrous work are shown in this table.

Tool	Use
Tape measure 	Used for measuring lengths such as laths, hessian and moulding work in lengths.
Pencil 	Used to mark out dimensions. It is impossible to work without one!
Timber or metal square 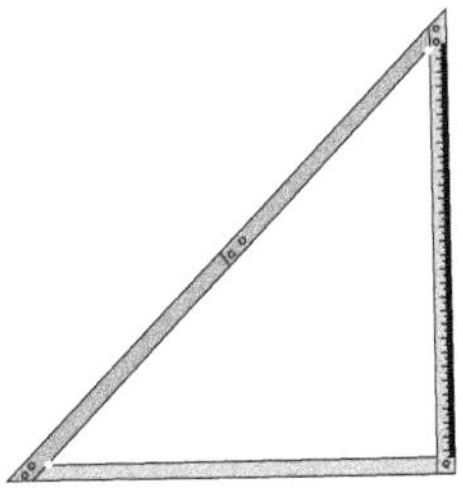	Used to square up reveals, soffits and beams.
Fine-toothed saw 	Used to cut plaster and timber for moulding.

Tool	Use
Rasp	Used to clean down and reduce the size of cornice and plasterboard.
Mitre box	Used to cut **mitres**.
Claw hammer	Used to fix (and remove) nails.
Battery-powered drill	Used to drive screws into a timber mould and for mechanically fixing fibrous work into plastic plugs or timber grounds.
Hand screwdriver	Can be used instead of a battery-powered drill/driver.
Tin snips	Used to trim zinc profiles and cut metal trims.

Tool	Use
Laser level 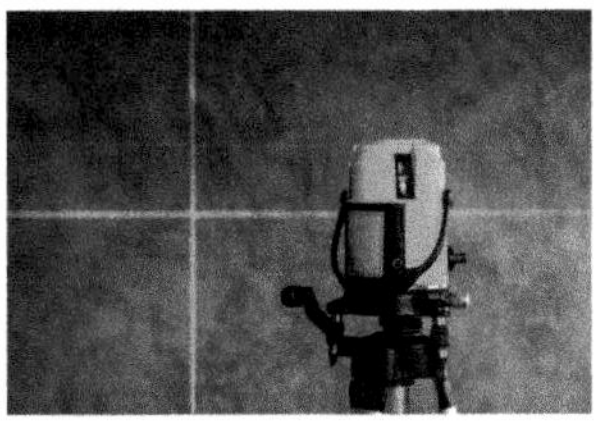	Used to determine a level or vertical line to work from.
Spirit level 	As with a laser level, used to determine a level or vertical line to work from.
Chalk line 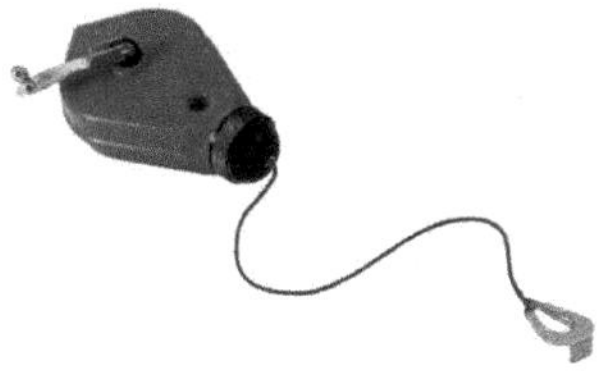	Essential in every fibrous plasterworker's kit: used to provide a line to work from.
Small tool 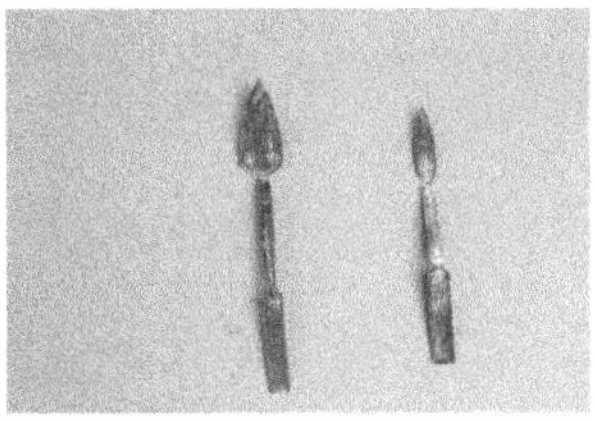	Essential for mitring joints.
Joint rule 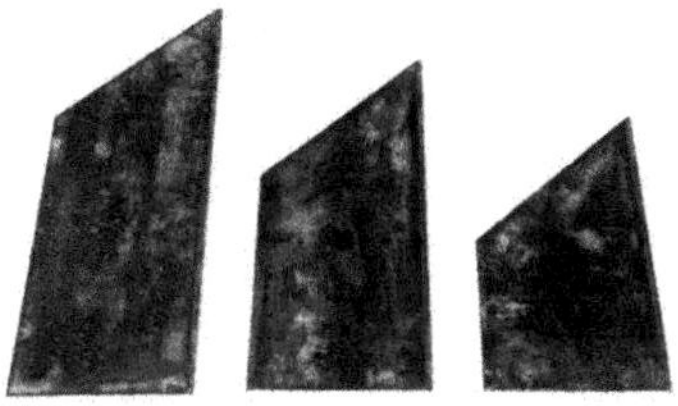	Used in conjunction with a small tool to form mitres.
Busk 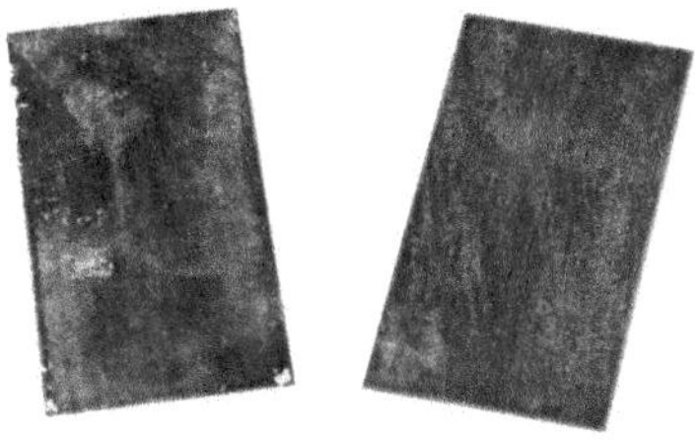	Used to remove blemishes from plasterwork.
Gauging trowel 	Used to mix small quantities of materials and apply materials to awkward areas.

KEY TERM

Mitre: a cut joint to an internal or external angle. The joint is then made good (the gaps are filled) with casting plaster or fixing adhesive, using a joint rule and a small tool.

Other equipment required includes:
- buckets and flexi bowls
- access equipment
- benches (if making fibrous work on site).

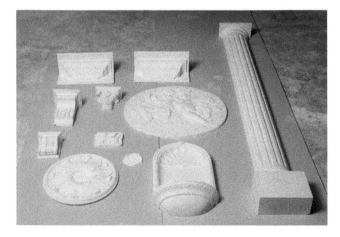

▲ Figure 6.49 Using a bench and a bowl to mix casting plaster

Cast fibrous plasterwork

The most common fibrous plaster casts are taken from a plaster, rubber or fibreglass reverse.

▲ Figure 6.50 Fibrous plasterwork

Setting up to cast from a plaster reverse

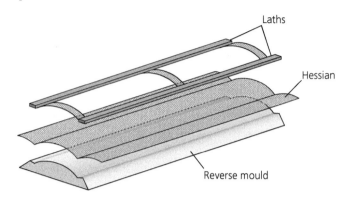

Laths

Hessian

Reverse mould

▲ Figure 6.51 A reverse

Materials and sequence of work

1. **Sealing the plaster reverse:** Use shellac to seal the plaster reverse mould. Apply many coats, leaving each coat to dry before applying the next.

▲ Figure 6.52 Sealing the plaster reverse with shellac

2. **Casting plaster and equipment:** While the shellac is drying, gather the necessary casting plaster, hessian cloth, timber lath, plasterer's grease or linseed oil, mixing bowls, brushes and hand tools.
3. **Hessian and lath:** Once the shellac has dried, lay the hessian along the length of the plaster reverse, ensuring that at least 50 mm extends at each end and both sides. (Remember that hessian comes in various sizes.)

- Cut some small pieces of hessian to use as cross braces; small pieces of timber lath can be used in conjunction with the hessian to improve strength.
- Cut the timber lath just short of each end of the plaster reverse; this is to allow for turning back the hessian cloth to the end and sides.
- Cut a small piece of lath for cleaning the plaster reverse **strike off** and set it aside for later.
- Soak the timber laths in water as this helps to reduce cracking in the fibrous cast.

4 **Release agent:** The most commonly used release agents are linseed oil and plasterer's grease, although other products are available. Apply a thin coat of your chosen release agent to the plaster reverse using either a brush or an oil-soaked hessian cloth.

5 **Bowls and buckets:** You will need at least three mixing bowls and two buckets. One of the buckets should contain clean water for mixing, the other water is for cleaning the splash brush during the casting.

▲ Figure 6.53 Cutting hessian

Producing the fibrous cast

Before you start:
- make sure everything you need is to hand and that the working area is clear of debris and clutter
- look down the length of the plaster reverse to check that you have applied a release agent to all of the plaster reverse mould
- check that the casting plaster is within its use-by date
- half-fill two mixing bowls with water, adding retarder to the second if you feel you will not have enough time to position hessian and timber lath reinforcement.

KEY TERM

Strike off: the built-up plaster area on the back of a cast that will come into contact with the background surface on the wall and ceiling when the plaster cornice is fitted in place.

INDUSTRY TIP

If you look along the length of the plaster reverse, you should see the oil glistening or a slight sheen. If any spots do not have this sheen, you have missed a bit!

This is the step-by-step procedure for using the two gauge system to produce a cast from a plaster reverse mould.

STEP 1 Cut the hessian and laths to the required length.

STEP 2 Apply grease to the face of the reverse mould.

STEP 3 Mix the firstings.

STEP 4 Apply the firstings to the reverse mould with a brush. Remove air from the cast by vibrating it, then leave it to pull in.

STEP 5 Wash the bowl as soon as possible, before the plaster sets.

STEP 6 Clean off the strike off before applying the seconds.

STEP 7 Mix the seconds and apply to the tacky firstings. Place the hessian in position, overhanging the strike offs.

STEP 8 Apply plaster over the hessian.

STEP 9 Bed the laths on the back of the cast.

STEP 10 Fold back the hessian/canvas over the laths to strengthen and reinforce the cast.

STEP 11 Build up the strike off with the remaining plaster, then leave to set.

STEP 12 Use a lath to form the strike off.

STEP 13 The cast will curl and lift slightly at both ends. However, do not remove it from the reverse until it has completely set.

INDUSTRY TIPS

Wet casting plaster has a slight sheen. When it starts to dull, the plaster is ready for the second application. Do not let the first application dry too much, as this will prevent the second application from adhering.

For larger casts, hessian and timber laths can be used to reinforce and strengthen the cast.

- Place the hessian and, if required, timber laths into the seconds bowl.
- Remove the excess plaster through your fingers and position the plaster-soaked hessian across the fibrous cast every 500–600 mm along its length.

ACTIVITY

Search online for 'plaster cast moulded cornice' to find out more about the techniques for casting from a plaster reverse.

Leave the plaster to set and then gently prise the fibrous cast away from the plaster reverse. Store it vertically on a hook or loop, if possible. If storing it flat, make sure the cast is kept completely flat otherwise it will distort.

If casting from a rubber reverse mould, silicone release agent must be applied instead of grease. The process and procedure for casting are otherwise the same.

Casting from a flexible mould

Casting from a flexible mould is similar to casting from a plaster reverse. The major difference is that there is no need to apply a release agent, because the hot and cold compounds are self-lubricating.

It is important to ensure that the flexible mould sits firmly on the workbench, as any distortion will show on the finished work. With larger flexible moulds, form a plaster or fibreglass case over the back of the flexible mould to ensure the flexible mould sits firmly on the bench.

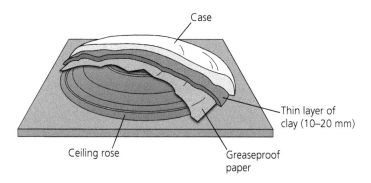

▲ Figure 6.54 Case mould

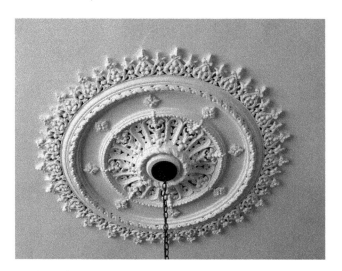

▲ Figure 6.55 A finished ceiling rose

Casting from a flexible ceiling rose mould

Follow these step-by-step instructions for the process of casting from a flexible ceiling rose mould.

STEP 1 Set up the workbench area: select the necessary casting plaster, hessian, lath, buckets, bowls and tools. Wash the flexible mould and pat it dry to remove surplus water. Place the flexible mould flat on the workbench. Check that the flexible mould is clean and dry.

STEP 2 Cut the hessian to overlap the ceiling rose by at least 50 mm all round, snipping around the outside to help prevent gathering. Cut extra hessian to cover the timber lath.

STEP 3 Cut the timber lath for the centre and the strike off.

STEP 4 Sift plaster into two plaster bowls to just under the water line, setting aside the second bowl without mixing.

STEP 5 Mix the first bowl of plaster to a creamy consistency and brush it into the decorative sections of the mould. Brush and then splash to cover to a depth of approximately 3–5 mm.

STEP 6 Clean off the strike off and allow the firstings to **pick up**.

STEP 7 Place a large piece of hessian on top of the firstings, then mix and pour a cup full of the seconds into the centre of the flexible ceiling mould.

STEP 8 Brush from the centre out towards the perimeter of the flexible mould. This action will minimise the build-up of air bubbles and ensure that the hessian adheres firmly to the firstings.

STEP 9 Work around the perimeter of the ceiling rose, tucking in the hessian overlap as you go, allowing for a 10 mm band around the perimeter of the ceiling rose.

STEP 10 With the large piece of hessian in position, brush both sides of the timber laths with plaster and lay them into position over the hessian. Dip the short pieces of hessian into the seconds, then tuck and lay the wet hessian over the laths, making sure there are no pockets of air.

STEP 11 Start to build one of the centre laths flush with the perimeter: this will allow the ceiling rose to fit flush with the ceiling. Splash the back of the ceiling rose and clean off the strike off for the last time.

STEP 12 Leave the plaster to set and then gently peel the rubber mould away from the plaster.

STEP 13 The finished ceiling rose. Remember to lay the plaster ceiling rose flat otherwise it will distort.

INDUSTRY TIP

Wet the laths before use. Sometimes thicker laths are left outside to absorb moisture.

▲ Figure 6.56 Wet the laths

Storage and delivery to site

Manufactured fibrous work should be stored in a dry, well-ventilated room, ideally with an ambient temperature of about 15°C. The important thing is to not let the room get too cold or too hot, otherwise the fibrous work could freeze, form mildew or dry too quickly.

Cornices are best stored upright; during casting, hessian loops can be incorporated into the casts so they can be hung vertically. If you decide to store a cornice flat, use plenty of **bearers** to promote drying; do not scrimp on bearers as wet fibrous work will warp if not supported correctly.

Other manufactured items should be stored flat to prevent any possible distortion.

All fibrous work must be protected from accidental damage caused by poor handling or poor transportation. Most cornice work is transported face to face and tied together. Bubble wrap is a good option as it can be used to completely wrap the work and will accommodate all shapes and sizes.

KEY TERMS

Pick up: when the materials start to stiffen.
Bearers: small blocks of wood placed between and underneath materials to keep them separate and promote drying.

HEALTH AND SAFETY

Always follow current health and safety legislation when moving and transporting materials:
- Health and Safety at Work Act (HASAWA) 1974
- Control of Substances Hazardous to Health (COSHH) Regulations 2002
- Manual Handling Operations Regulations 1992
- Personal Protective Equipment (PPE) at Work Regulations 1992.

You should also pay attention to the Environment Agency's requirement that plaster waste be disposed of separately and make sure you plan for deliveries and storage on site.

For more information about health and safety legislation, refer back to Chapter 2.

⑤ FIX AND FINISH PLASTER MOULDINGS

How to fix fibrous plasterwork

The first task is to consult the contract schedule and contract drawings to determine precisely where the fibrous items are to be fixed. You will also need to refer to the contract specification to determine the specified fixing methods and materials. Remember that:
- the contract schedule identifies the fibrous items and where they are to be fixed
- the contract drawings identify the fibrous items in detail, with fixing dimensions
- the contract specification identifies fixing methods and materials.

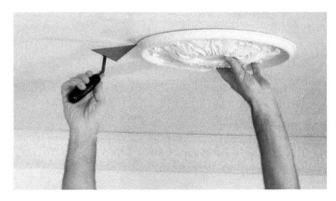

▲ Figure 6.57 Fixing a ceiling rose

Fixings

Most fibrous work is fixed into position using adhesive, although larger, heavier items might require additional support using mechanical methods. The length of the fixing will depend on the thickness of the fibrous item being fixed and the depth of the background it is being fixed to.

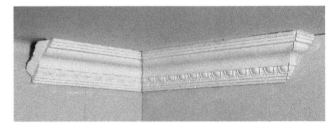

▲ Figure 6.58 Fixing fibrous plasterwork

Suitable fixings include:
- adhesive
- nails and screws
- wire and wad.

Adhesive

Coving/fibrous cornice adhesive is available ready-mixed or in powdered form. It can be purchased in 1 kg, 5 kg, 10 kg, 15 kg, 20 kg and 25 kg bags.
- Ready-mixed adhesive has a slower setting time as it is air setting.
- Powdered adhesives have specific setting times that range from 45 to 120 minutes, depending on the manufacturer.

The adhesives are used for jointing internal and external mitres. The more expensive adhesives incorporate a PVA that improves adhesion to the background.

Nails and screws

Nails must be rust-proof, such as galvanised, aluminium or zinc-plated nails. Those with annular shanks offer better fixing as they grip the timber better. Nails usually come in 25 mm, 30 mm, 40 mm and 50 mm length sizes.

Nails are a less popular fixing option as they tend to lose their fixing strength over time and can result in nail-popping with plasterboarded ceilings.

In contrast, screws offer a more secure fixing and in most cases are cheaper to buy than galvanised or zinc-plated nails.

Drywall screws are commonly used because they are phosphate or zinc-plated, which makes them rust-proof. They are available in most sizes from 25 to 75 mm, which covers the majority of fixing situations.

When fixing into masonry, plastic plugs are required to properly secure the screw fixing so that it can carry the weight of the fibrous work. The plastic strip that joins the plugs contains guidance on the required size of the screw and the diameter of the masonry drill bit to drill the hole in the masonry.

The size and weight of the cornice dictate how much fixing is required. Heavy cornice should be mechanically fixed with a plastic plug and screws every 300 mm, while lighter cornice can be mechanically fixed every 600 mm, also using a suitable fixing adhesive.

▲ Figure 6.59 Annular nails

ACTIVITY

Search the internet to find out:
1 what nail-popping means
2 the cost of:
- 1 kg of 50 mm plasterboard nails
- 500 50 mm drywall screws
- a box of 1000 coarse phosphate 50 mm-long screws.

▲ Figure 6.60 Drywall screws

▲ Figure 6.61 Plastic plug

Wire and wad

Wire and wad is a traditional method of hanging fibrous slabs from joists and steel beams. Basic wadding involves soaking hessian in casting plaster to fix or joint fibrous work together or to a background. To improve the strength, galvanised wire is tied to the fibrous work and to a fixing point and then wrapped with plaster-soaked hessian.

Galvanised wire is used because it is easy to work with and does not rust. The choice of wire thickness depends on the size and weight of the fibrous slab, but typically it will be 1.6–5 mm thick.

As soon as the fibrous slab has been levelled and secured, plaster-soaked hessian is wrapped around the wire to form a strong support mechanism. It is important to make adjustments to the fibrous slabs before applying the plaster wads, because as soon as the plaster wads set, no further adjustments can be made.

Marking out and preparation

Take the following measures to mark out and prepare for installing fibrous plasterwork.

1 Lay protective sheeting on the floor area and select suitable access equipment.
2 Check that the background and the item to be fixed are both free from dust.
3 Determine the projection and depth of the coving or cornice.
4 Using a level and chalk line, mark out the position of the fibrous item to be fixed.
5 Provide a mechanical key by lightly scoring the background with a craft knife or a scutching hammer.

6 Adjust the suction of the background and the fibrous item using clean water or a weak solution of water and PVA.
7 Have ready some temporary support such as nails or props to hold the fibrous item in position until the adhesive has set. (If permanent mechanical fixings are to be used in conjunction with adhesives, these temporary supports are not so important.)

Fixing cornices

Before starting the work, have a final check of the drawing and manufacturer's information to determine the projection and depth of the cornice.

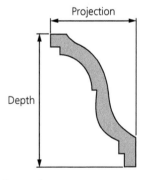

▲ Figure 6.62 Projection and depth of a cornice

Although the following text refers to cornices, the same steps should be followed for fixing coving.

The walls and ceilings should be plumb, level and true. If they run out slightly, it is better to have straight projection lines, as these lines are more noticeable when looking at coving or cornice work.

Once you are ready, follow these step-by-step instructions.

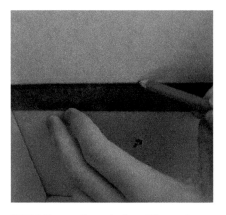

STEP 1 Measure the projection of the cornice on the ceiling line (in this case, 100 mm).

STEP 2 Measure the depth of the cornice on the wall (in this case, 100 mm).

STEP 3 Snap chalk lines to the ceiling projection and depth marks to indicate the position of the cornice on the ceiling and wall.

STEP 4 Place the cornice in position between the set lines and mark the cornice to indicate the mitre cut.

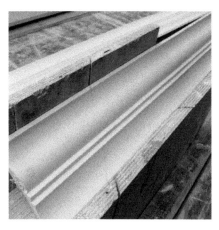

STEP 5 Position the cornice in the mitre box, making sure the ceiling line of the cornice lies to the base of the mitre box.

STEP 6 Use a fine tooth saw to cut the cornice to your previously set mitre marks, making sure you cut to the front and not from the rear.

STEP 7 Cut a stop end on the other side. This will be the same cut as an external mitre.

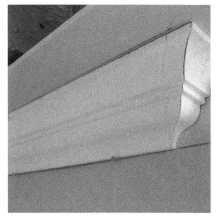

STEP 8 Fix the stop end in position.

STEP 9 Repeat the process when cutting the next length which starts with a left internal mitre cut, followed by the opposite external mitre on the opposite side.

STEP 10 Check that this fits and tack in place.

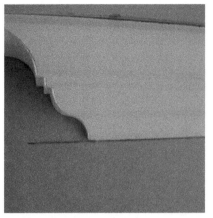

STEP 11 Cut the remaining length which has an internal right mitre and on the opposite side a return stop end.

STEP 12 Now you have cut all the cornices and checked their position, you need to remove them and prepare them for fixing.

STEP 13 Prepare the cornice by sealing the strike off with diluted PVA; this will control the suction.

STEP 14 Mix the cornice adhesive to a creamy paste and apply adhesive to the strike offs with a gauging trowel.

STEP 15 Position the cornice directly in place and firmly squeeze; this will cause excess adhesive to spread out.

STEP 16 Use a busk to remove the excess adhesive.

STEP 17 Clean the wall and ceiling line using a brush and clean water, to remove any surplus adhesive.

STEP 18 Continue to fit the remaining lengths and both stop ends, making sure that all the moulding members line in with no steps. Moulding members should be sharp and in line with no misses, no build-up of plaster, no blemishes and no **chattering marks** on the moulding surface.

KEY TERM

Chattering marks: marks caused by chattering (excessive expansion of casting plaster as the mould is run along the bench when producing moulding).

INDUSTRY TIP

When cutting a mitre, always have the ceiling line of the cornice on the bench with the face of the cornice towards you.

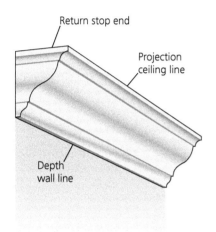

▲ Figure 6.63 Cornice in position

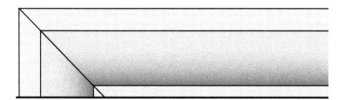

▲ Figure 6.64 Plan view of stop end

Cutting splayed joints requires a different technique, with the projection and deep chalk lines marked out.

- For an internal splayed joint: cut the cornice to the wall length and hold into position. Where the ceiling line chalk lines cross, mark the cornice and cut to the end of the cornice meeting the wall line.
- For an external splayed angle: cut the cornice to the wall line and allow for the length of the projection. Again, hold the cornice in position and, where the chalk lines cross, make the cut to the end of the cornice on the wall line.

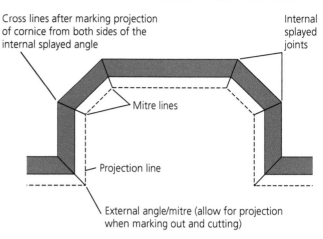

▲ Figure 6.65 Plan of bay window

IMPROVE YOUR MATHS

1 The quantity surveyor suggests that a single plasterer will fix a 3 m length of fibrous cornice, including all jointing, in 40 minutes. How long will it take the plasterer to fix all the fibrous cornice in a living room with a perimeter of 38 m?

2 The fibrous cornice for the living room has been manufactured at a cost of £26.35 per 3 m length. How much will it cost to manufacture?

IMPROVE YOUR ENGLISH

1 A Victorian style ceiling rose is to be fixed to a single central light. Write a specification on how best to fix the fibrous ceiling rose to the ceiling.

2 Write a method statement on fixing a cornice to a living room.

INDUSTRY TIPS

If the fibrous cornice is dry, adjust the suction by applying a weak solution of PVA. If you apply weak PVA to joints and mitres, it improves adhesion and helps prevent hairline cracks.

- Larger mitre gaps can be filled out with hessian soaked in plaster.
- Number each panel section and the background in sequence. This saves time and ensures that all the items fit.

Panel and dado moulds

Panel and dado moulds are usually smaller and lighter than cornice moulds and only require adhesive for fixing. You will still require temporary support to hold mouldings in position.

- The background should be clean and free from debris.
- Having marked out fixing positions with a level and chalk line, lightly score the background and the back of the moulding to provide a mechanical key.
- Test suction of both the background and the moulding with a splash of water and adjust accordingly with water or a weak solution of water and PVA.
- As with the cornice, the moulding must sit firmly and squarely in the mitre box. Try not to cut the mitre exactly: a small mitre gap of 5–8 mm allows adhesive to be pushed in and forms a stronger mitre.
- Some panel moulds feature a quadrant mould running into a straight length, similar to cutting a splayed joint to a cornice. A bit of skill is required to produce a mitre as these moulds cannot be cut in a mitre block. To overcome this problem, outline the shape of the panel mould on the wall or ceiling. Make mitre lines longer so that you can see where to mark the panel moulding for cutting. Draw a line through the outer and inner lines and extend the lines by about 5 mm to allow for the mitre joint.
- Cut all of the panel mould lengths and hold them in position to check that they fit before mixing any adhesive.
- Mix a small amount of adhesive to a thick, creamy paste and apply with a gauging trowel to a thickness of about 3 mm to the back of the moulding. Press firmly to the outer or inner outline and remove any surplus materials with a joint rule or a damp sponge.
- Work all mitres with a small tool and joint rule.
- Remove temporary support when appropriate and make good as necessary.

Beam cases

Beam cases can be manufactured in one unit, i.e. upstand cheeks and a soffit, or they can be manufactured in three sections, i.e. two upstand cheeks and one soffit. Alternatively, timber laths can be fixed to the steel beam or concrete beam with suitable fixings to allow the beam case to be fastened to them using suitable screws.

The method of fixing a beam case depends on the beam it is surrounding. A structural beam will be formed from concrete or a steel joist; imitation beam background may be formed with timber or metal studding. Therefore beam cases are fixed using a combination of adhesive and mechanical fixing.

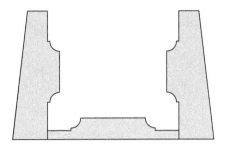

▲ Figure 6.66 Beam case

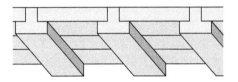

▲ Figure 6.67 Bare structural concrete beam floor

▲ Figure 6.68 Beam cases in position

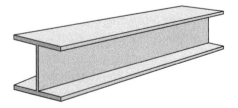

▲ Figure 6.69 Steel beam

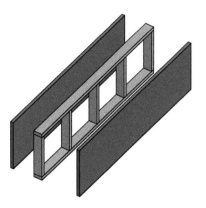

▲ Figure 6.70 Built-up timber beam

Fibrous slabs

The method for fixing fibrous slabs depends on the type of ceiling, whether it is:
- timber joist or metal stud
- suspended.

Timber joist or metal stud ceiling

When fixing to a timber joist background, the finished ceiling height datum line will have been formed by a carpenter. The fibrous slabs should have been manufactured to fit to the joists' centres, with slabs being mechanically fixed to the ceiling joists using drywall screws or galvanised nails into timber joists.
- The starting point will vary, depending on the size and layout of the room.
- During fixing, drill pilot holes through the slab at fixing points to minimise the risk of cracking the slab. The pilot holes are usually just smaller in diameter than the fixing screw or nail.
- Think about the weight of the fibrous slabs before attempting to hold and fix them in position. Two people working together will make the job easier. If working on your own, a **strut** or **prop** can be used.
- You must check slabs for alignment as the work proceeds, similar to plasterboarding a ceiling.

KEY TERM

Strut/prop: a telescopic pole with pads on each end; the pole is adjusted to hold an item above your head, just like an extra pair of hands. A useful piece of equipment when working on your own.

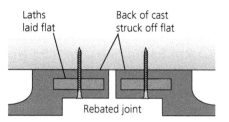

▲ Figure 6.71 Fixing a fibrous slab with drywall screws

Suspended ceiling

Suspended ceilings are used to lower a ceiling's height and to conceal services. The architect will specify a datum point from the finished floor level or a datum level. A metal grid is formed which hangs from the concrete floor slab or structural beams. Fibrous slabs are usually wired and wadded into position.

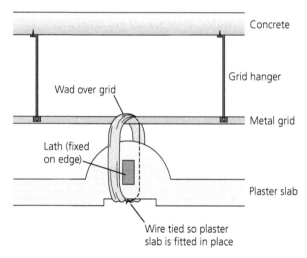

▲ Figure 6.72 Detail section of fibrous slabs fixed with wire and wad

Sequence of work

1 Set out ceiling grid centres.
2 Pull the string line tight along the first line of slabs.
3 Drill holes each side of the lath diagonally at 70–80 mm centres to line up with the fixing points of the grid.
4 Form slotted grooves or rebates between the holes below the surface of the slabs.
5 Working in pairs with one person supporting the slab, thread a suitable length of 18 g wire through the slab and twist the wires.
6 Adjust the wires if required, using wire cutters, until the slab is level and in line.

7 Wrap the wires with plaster-soaked hessian to form a solid plaster fixing.

8 Repeat the steps above for the other slabs in the line.

9 Reposition the string for the next row of slabs and repeat the process until the ceiling is complete.

IMPROVE YOUR MATHS

The local builders' merchant's delivery van has broken down. Your manager asks you to collect 30 bags of casting plaster weighing 25 kg.

How many trips will you need to make to collect the materials if your van's maximum carrying capacity is 600 kg?

Jointing fibrous slabs

- Wet the joints with a weak solution of water and PVA.
- Push hessian soaked in plaster into the rebated joint, making sure the hessian does not protrude.
- As the hessian stiffens in the rebate, mix a small quantity of casting plaster and work it into the joint with a small trowel.
- Finish the joint with a joint rule or busk; the hessian will strengthen the joint and reduce the risk of cracking.

▲ Figure 6.73 Fibrous slabs

Producing a cast from a reverse mould

Steps 1–10 show how to produce a cast from a reverse mould and fix it to a wall with an internal and external corner.

1 Produce two lengths of cornice by casting from a prepared reverse mould you have produced during your training.

2 Mark out the projection and depth measurements on a background which includes one internal mitre, one external mitre and a stop end.

3 Key the background by scoring and then seal the surface with diluted PVA to control the suction.

4 Using a saw, cut the cast mouldings to the required length and correct to mitred cuts.

5 Fix nails in the wall and ceiling and position the cast dry to ensure everything fits accurately.

6 Prepare and seal the rear of the casts for fixing and mix the necessary adhesive.

7 Apply adhesive to the cast and bed in correct position.

8 Clean wall and ceiling lines and repeat process for remaining casts.

9 Ensure mitres are in line and intersecting after positioning.

10 Make good all joints by stopping in, ensuring mitres are sharp and members are lining through.

CASE STUDY

Kwame and Sam arrive on site to fix the cornice casts they produced in the workshop.

Kwame fetched the clean dust sheets from the van and laid them in the room where the cornice was to be fitted. Sam got the tools and equipment out of the van and set up a working area.

Reading the specification, Sam told Kwame to check the projection and depth of the cornice and the fixing method. Using a working platform to reach the ceiling line, Kwame used a tape and measured out the projection across the ceiling and the depth down the wall from the ceiling line. From each corner, he snapped chalk lines around the perimeter.

Wearing the correct PPE, they carefully created a mechanical key by lightly scoring the plasterwork with a utility knife, sweeping the dust away as they proceeded. Kwame then mixed some PVA to a 1:5 ratio to seal the plaster surface. They used Fiberfix adhesive as the coving is lightweight.

Sam instructed Kwame to fix in nails just below the chalk line on the wall, while he measured and cut enough cornice for the first wall. Kwame mixed a small amount of adhesive to fix these lengths. Sam applied adhesive to the fixing edge of the cornice with a gauging trowel, ensuring a liberal application was applied. Working together, Kwame and Sam carefully lifted the cornice into position and squeezed it against the wall and ceiling chalk lines. Sam fixed a few nails along the ceiling line to temporarily hold the cornice in position and then removed the surplus adhesive with a joint rule and cleaned the surface with clean water and a small brush.

They worked in sequence around the room until all of the cornices were fixed in position. Kwame removed the temporary support nails as the adhesive started to set, while Sam mixed a small amount of casting plaster to make good and stop in the internal mitres using a joint rule, small tool and small brush. Kwame used a joint rule to fill the holes where the nails temporarily held the cornice in position, then wiped the ceiling and wall joints with a moist sponge to clear away any dirty plaster marks.

They both walked around the room, checking that the cornice was fixed and clean. Satisfied, Sam cleaned all the tools and equipment while Kwame carefully removed the dust sheets, shaking them in the back of the van to minimise mess in the client's property.

As soon as the room was clean and tidy, Sam asked the client to look at the complete work. The client was pleased with the standard of work and complimented Kwame and Sam on the cleanliness and efficiency of their work.

Test your knowledge

1   What can be used to take a copy of a cornice?

   A   squeeze

   B   slurry

   C   clasp

   D   clutch

2   What term is used to measure how much the cornice protrudes from the ceiling?

   A   projection

   B   volume

   C   length

   D   depth

3   Which stage does 'running down' refer to when producing a positive mould?

   A   when preparing to run

   B   after completing the first run

   C   during the building up process

   D   when completing the final run

4   What is the advantage of using hot melt compounds for producing fibrous plaster mouldings?

   A   casts do not need reinforcement

   B   you can produce multiple casts

   C   casts will dry quickly, saving time

   D   compounds can be remelted and reused, saving costs

5   What is the advantage of using cold pour compounds for producing fibrous plaster mouldings?

   A   melting equipment is not required

   B   you can produce multiple casts

   C   casts will dry quickly, saving time

   D   compounds can be remelted and used again

6   What material is used as a release agent?

   A   paraffin

   B   clay

   C   grease

   D   shellac

7   What can occur if completed plaster casts are not stored correctly?

   A   dusting

   B   distortion

   C   warping

   D   shrinkage

8   What is fixed to a bench to prevent the reverse mould from moving during the running process?

   A   screw

   B   scrim

   C   rule

   D   nail

9   Firstings and secondings is a method used for producing plaster casts by using how many gauges?

   A   one gauge

   B   two gauges

   C   three gauges

   D   multiple gauging

10 What is fixed to the bench to make sure the running mould runs in line?

   A   box rule

   B   timber rule

   C   scale rule

   D   measuring rule

11 What material is used to seal plaster reverse moulds?

   A   shellac

   B   grease

   C   chalk

   D   size

12 What material is used to reinforce fibrous plaster casts?

   A   clay

   B   tallow

   C   laths

   D   hessian

13 A contract requires 122 m of cornice to be fixed. How many 3 m lengths are required? Add 10% for wastage and round up your answer.

A 44

B 45

C 46

D 47

14 How many cornices measuring 3 m in length are required for a room measuring 8 m by 5.3 m?

A 7

B 8

C 9

D 10

15 When casting a length of cornice, when is reinforcement applied?

A when applying the firstings

B when applying the seconds

C after the seconds have set

D after the firstings have set

16 Why is shellac applied to plaster reverse moulds?

A to stabilise dry dust

B to protect the surface

C to seal the surface

D as a release agent

17 What method is best used for carrying lengths of fibrous plaster casts?

A on their edge in pairs

B flat-faced, end to end

C over your shoulder

D upright, individually

18 When cutting mitres, where is the projection of the mould placed in the mitre box?

A to the left side of the box

B on the floor of the box

C on the upstand of the box

D to the right side of the box

19 What type of fixings are preferred when installing cornice?

A rustproof

B flexible

C expandable

D compressible

20 When fixing cornice, what do chalk lines indicate?

A depth and projection

B thickness of cornice

C length and dimensions

D height of room from the floor

21 What is used as a temporary fix when installing fibrous plasterwork?

A sink screws

B dry wall straps

C plastic plugs

D steel nails

22 What is the **main** advantage of fibrous plasterwork, in addition to the fact that it is strong and light?

A It can be prefabricated.

B It needs no decoration.

C It needs no mechanical fixings.

D It can be load-bearing.

23 What should be applied to the reverse mould before casting takes place?

A shellac

B size

C grease

D chalk

24 What can be added to slow down the set of casting plaster?

A tallow

B alum

C size

D soda

CHAPTER 7

METAL FRAME SYSTEMS

INTRODUCTION

In this chapter we look at three different types of metal frame system used in today's construction industry. By reading this chapter you will gain knowledge of:

1 the benefits of using metal frame systems
2 information sources for using metal frame systems
3 how to set out metal frame systems
4 how to install metal stud partitions
5 how to install metal wall lining systems
6 how to install metal furring ceiling systems.

The table below shows how the main headings in this chapter cover the learning outcomes for each qualification specification.

Chapter section	Level 1 Diploma in Plastering (6708-13)	Level 2 Diploma in Plastering (6708-23) Unit 222	Level 2 Technical Certificate in Plastering (7908-20) Units 207, 208, 209
Benefits of using metal frame systems	n/a		Topic 1.1
Information sources for using metal frame systems	n/a	Learning outcomes 1.1, 1.2, 1.3, 1.4	Topics 1.2, 1.3
Set out metal frame systems	n/a		Topics 2.1, 2.3
Metal stud partitions	n/a		Unit 209 Topics 2.2, 3.1, 3.2, 3.3, 3.4
Metal wall lining systems	n/a		Unit 207 Topics 2.2, 3.1, 3.2, 3.3, 3.4
Metal furring ceiling system	n/a		Unit 208 Topics 2.2, 3.1, 3.2, 3.3, 3.4

1 BENEFITS OF USING METAL FRAME SYSTEMS

Using metal to build has significant benefits compared to traditional timber methods. Plasterers use metal to install partitions, linings and ceiling systems. Metal frames or linings can be built and installed much more quickly than timber, which can be bulky and heavy, and requires the use of power tools for cutting and installing. Timber can also rot, warp and shrink, causing further problems for plastering surfaces.

Other advantages of using metal:
- It is non-flammable.
- It is easy to transport, store and handle.
- It is easy to cut and install.
- It can include insulation which will improve the thermal properties of the building.

- It is easy to accommodate mechanical and electrical services such as water, electricity and telecommunications.
- Metal substrates can be installed, which help to control noise and can also add **fire proofing**.

> **KEY TERM**
>
> **Fire proofing:** installed to increase the fire rating of the metal frame.

In this chapter we will look at three types of metal system:

1 **Metal stud partitions** are made up of a track channel profile with incorporated metal studs to form a frame. They are used to divide large open-plan areas into rooms in private, residential and commercial buildings.
2 **Metal wall linings** provide a true lineable surface for misshaped walls. The lining is fixed to a solid background surface that might require straightening or upgrading.
3 **Metal ceiling linings** provide a metal grid background to form accurate level ceiling systems in old and modern buildings. Using steel will increase the fire rating.

Each system has its own installation guidelines and provides a backing that can easily be covered with plasterboard, which can be either plastered or taped and jointed.

② INFORMATION SOURCES FOR USING METAL FRAME SYSTEMS

Drawings

When installing metal frame structures, it is important to read the drawings before you start as they will provide you with:

- the setting out dimensions and position of the frame
- the components used with the installation system.

Floor and reflective drawings provide information on layout.

Detail drawings provide specific information on design, such as deflection heads (see page 288).

Specifications

These documents provide information about the type of metal section, component or fixing to be used. You can use the specification when ordering materials as it provides specific information about the materials to be used.

Specifications state the necessary standards of workmanship that need to be adhered to when installing metal-framed systems.

Manufacturer's information

The manufacturer provides information to enable you to install the metal system in line with their instructions. This document is a useful guide to prevent poor workmanship. Consumers can typically find troubleshooting questions and answers in the product instructions if problems arise.

Building regulations

Building regulations are standards intended to protect people's safety, health and welfare in and around built environments. These are the minimum standards for design and are supported by approved documents; for example, setting standards for energy performance, resistance to the passage of sound and sound insulation, or fire safety within and around buildings.

The local authority's building control department is responsible for checking that the building work meets building regulations standards.

③ SET OUT METAL FRAME SYSTEMS

Before installing the system, you will need to make sure the surface background is fully prepared.

Preparing surface backgrounds

Before installing the system you will need to do the following checks:

- **Analysis**: before installing partitions, you need to assess background substrates for strength, straightness and mechanical fixing process. This will establish the type of anchor points and fixings to be used.
- **Type of surface**: there are a number of different types of surface to which the system may be installed, including concrete, beam and block, brick, block, timber and metal. Different surfaces require different fixing centres.
- **Condition**: you will need to check the strength of substrate to receive the metal-framed system.

Methods of setting out

When setting out you will need to consider:
- floor and ceiling lines and right angles
- doorways and openings
- fixing points and centres
- services.

For more details, see the sections on installing the different types of system below.

4 METAL STUD PARTITIONS

There are several types of metal stud which can be used when erecting metal stud partitions:

- Metal studs are incorporated into a ceiling and wall channel to form the partition frame, as shown in Figure 7.1.
- Metal studs incorporated in the track provide vertical support in wall framing. These are called **C studs** and are available in a range of widths, lengths and **gauges** depending on requirements for strength height, impact resistance, sound insulation and any need for **service channels**.

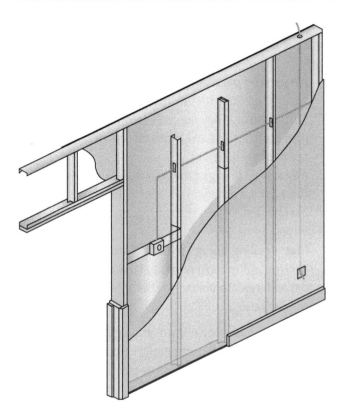

▲ Figure 7.1 Metal stud partitions

KEY TERMS

C stud: named after the profile shape of the metal stud.

Gauge: the term used to indicate the metal's thickness.

Service channels: gaps or channels incorporated in the studs to allow for cables and piping.

Door opening

▲ Figure 7.2 Metal stud partitions showing door opening

Metal stud components

Component	Use
Track 	Fixed to the floor and ceiling to receive the upright metal studs. Also used to form heads and sills for openings.
C stud 	Metal studs are incorporated between the floor and ceiling track to produce the metal-framed partition wall surface that receives the installation of plasterboard.
I stud 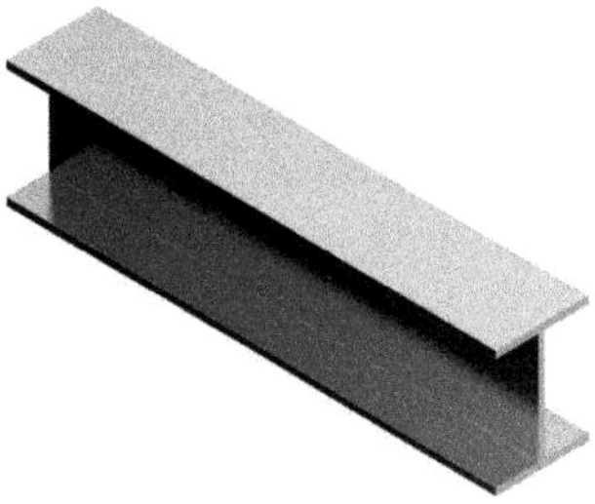	Used when increased strength and impact resistance are required.
Acoustic stud 	Manufactured for increased acoustic performance.
Wafer head drywall screw 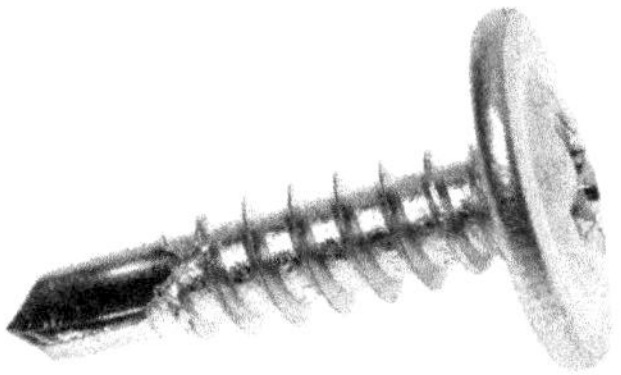	Used to fix together metal components. These are corrosion-resistant self-tapping steel screws.
Wafer head jack-point screw 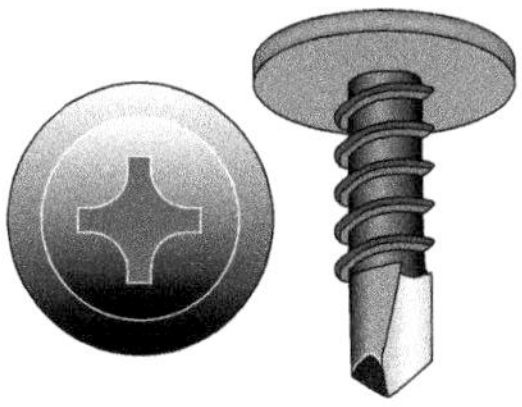	Used when the depth of metal is greater than 0.8 mm thick. Fixing metal components to masonry can be carried out using mechanical fixings with screws and plugs or a cartridge-operated tool.

When forming openings for doors, the track is returned a minimum of 300 mm up the stud from the floor and securely fixed by crimping or screwing using wafer head screws. These returns are referred to as **jambs**. To form the head detail, the track is cut leaving a minimum of 150 mm to be returned down the stud. Once the door opening head track is positioned to meet the height of the door frame, dimensions the returns can be crimped or screwed in place.

▲ Figure 7.3 A PH2 drill bit

ACTIVITY

Search online for different types of mechanical fixing used for installing and fixing metal components to masonry. Look for different manufacturers and systems to broaden your understanding of this topic. Some websites that might be useful include:

- www.british-gypsum.com/literature/white-book/partitions
- www.knauf.co.uk
- www.siniat.co.uk/en/knowledge-centre

Manufacturers provide installation guidelines and videos on how to build systems step by step.

INDUSTRY TIPS

- If you use perforated metal tack and stud, a screw fixing will be able to penetrate both metals without slipping.
- Screws used for interior systems have been designed to be installed using a PH2 drill bit.

KEY TERMS

Jambs: when the track is cut and returned to form a door or window opening in a partition.

Datum line/point: a point or line from which measurements are taken when setting out.

Tools and equipment used for installing and constructing metal stud partitions

Tool/equipment	Use
Tape measure	Used to set out and mark dimensions for cutting and installing metal track and studs.
Laser level	Can accurately project a red or green laser beam from fixed **datum lines/points** when establishing straight lines, right angles and levels for setting out floor and walls to receive metal track and studs.
Tin snips	Used to cut metal components to the required length.

Tool/equipment	Use
Chop saw 	As it has a metal blade, a chop saw is used when there is greater demand for cutting metal components; for example, in large-scale construction work.
SDS hammer drill 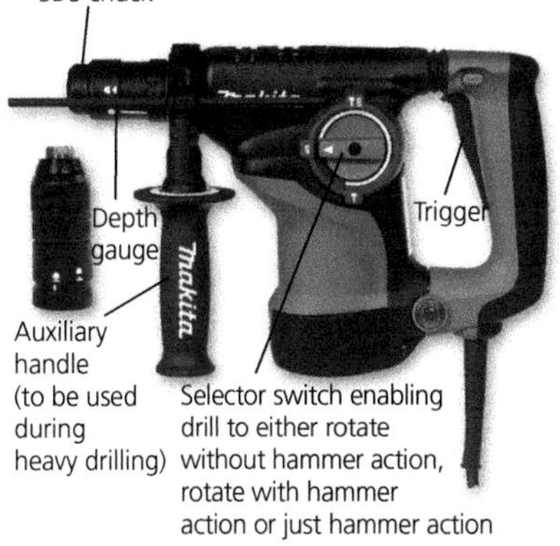	Used for drilling holes in masonry backgrounds when using mechanical fixings.
Portable screwdriver 	Used for fixing screws.
Crimper 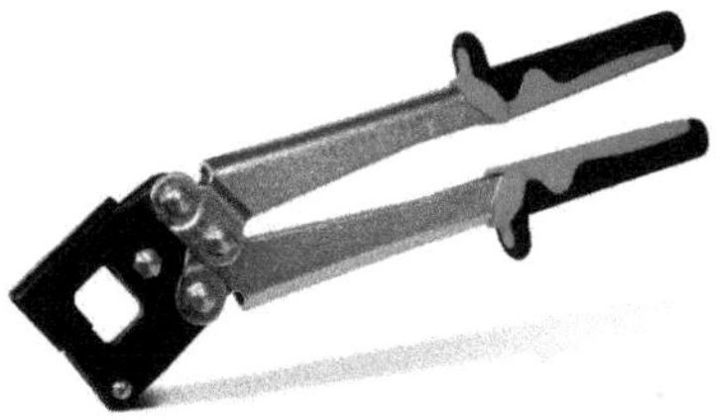	A hand tool that punctures the metal track and stud together without the need to use wafer head screws.
Clamps 	Used to provide a temporary fixing when preparing and positioning metal components before final installation.

Tool/equipment	Use
Magnetic levels	Useful when levelling and plumbing metal before fixing in position.
Square	Used to mark right angles.
Impact drill driver	Used when additional power and torque are required to drive a screw without damaging the screw head.

IMPROVE YOUR MATHS

Research the costs of the items in the table above. Calculate the cost of each item and then work out the total cost of the tools.

When installing metal-framed systems, several types of access equipment are used to ensure that work is carried out safely at height.

Type of access equipment	Use
Hop-up	Provides a low level platform for a single person.

Type of access equipment	Use
Step ladder 	Generally used for light work or accessing platforms.
Tower scaffold 	Suitable for one or more people. It has a working height greater than a hop-up or podium.
Podium	A suitable working platform for one person which can be moved around the working area with ease.
Scissor lift 	A motorised mechanical lifting aid that provides a low or high level working platform.

Deflection heads are used within metal stud partitions when there is a need to allow for movement (up, down or both) within the structure at the *head* of a partition. This movement (or *deflection*) is created by loads on the floor or roof above.

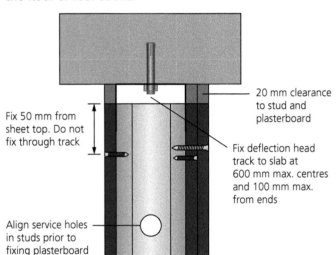

Fix 50 mm from sheet top. Do not fix through track

20 mm clearance to stud and plasterboard

Fix deflection head track to slab at 600 mm max. centres and 100 mm max. from ends

Align service holes in studs prior to fixing plasterboard

▲ Figure 7.4 Deflection heads.

Install metal stud partitioning systems

Step-by-step procedure for installing a metal stud partition with a door opening

STEP 1 Measure and mark out on the floor and ceiling, making sure the floor marks are plumbed up to the ceiling. The most accurate way to do this is to use a laser level.

STEP 2 Measure, cut and fix floor track, allowing for the door opening. Ensure the track is returned upwards along the stud a minimum 150 mm and fixed securely. You can use mechanical fixings when installing the track to concrete floors and dry wall screws to fix the ceiling track to the timber joists.

STEP 3 Measure and cut the length of track for the ceiling and fix in position. This is now ready to receive the studs. Measure and cut stud uprights 10 mm shorter than required to enable ease of installation.

STEP 4 Position the studs at 600 mm centres. Where the stud meets the masonry, install the metal with mechanical fixings at 600 mm centres.

STEP 5 Fix each stud in the correct position, using wafer head screws.

STEP 6 Form the door head by measuring and cutting a length of track. Position and fix in place, ensuring that it supports the stud.

STEP 7 To produce a strong door lining with good fixing points, install timber fillets that are inserted into the door opening metal studs. Cut a sheet of plywood and fix it between the studs to create a fixing for a radiator. You could also consider this if there will be cupboards fixed to the wall.

The framework is now ready to receive the plasterboard.

INDUSTRY TIP
When forming a right angle stud wall, leave a gap to fit the plasterboard.

ACTIVITY

1 Search the internet for different manufacturers of metal partition systems.
2 Find the dimensions (the length and width) of metal stud and track components.

Gap for drywall   Loose stud   Drywall screw

▲ Figure 7.5 Metal stud framing

Fixing details when forming a return using metal stud

Dviding walls are used to divide large spaces into different room layouts. When installing such walls, you will be required to set out and install metal studwork that contains returns and right angles. The drawings below show the positions of plasterboard and studs to ensure the appropriate fixing methods are used and to avoid poor workmanship.

KEY TERM

Dividing wall: a wall that separates areas; for example, for framing out bedrooms and bathrooms. This will not be a load-bearing wall.

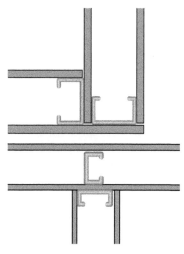

▲ Figure 7.6 Layout of a metal stud partition corner (top) and a metal stud partition containing a right angle (bottom)

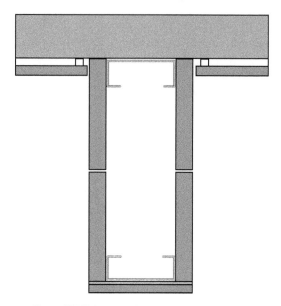

▲ Figure 7.7 Fixing metal stud to masonry substructure

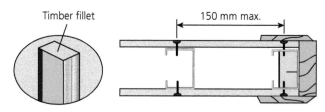

Timber fillet   150 mm max.

▲ Figure 7.8 Inserting timber fillets in metal studs to receive door frames

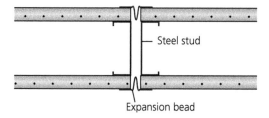

Steel stud

Expansion bead

▲ Figure 7.9 **Movement joints** in metal stud work

KEY TERM

Movement joints: these allow surface and components to move slightly under expansion and contraction in buildings.

Sound proofing: installed to reduce noise transfer through metal systems.

Intumescent sealants

Intumescent sealants are designed to stop the spread of fire and smoke in interior systems installation.

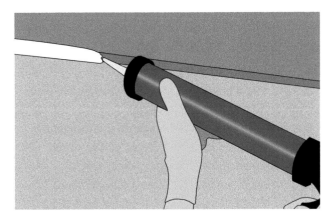

▲ Figure 7.10 Intumescent sealant

Acoustic sealants

Acoustic sealants are a form of **sound proofing**. They are designed to reduce sound transmission and are used with interior systems installation.

▲ Figure 7.11 Acoustic sealant

IMPROVE YOUR MATHS

Calculate how many C studs you would need to divide a room measuring 6 m in length if the studs are fixed every 600 mm.

⑤ METAL WALL LINING SYSTEMS

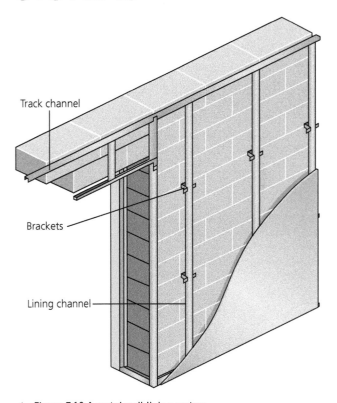

▲ Figure 7.12 A metal wall lining system

A metal wall lining system is a lightweight, non-load-bearing dry lining which is erected independently of the external wall construction.

Wall linings are generally used when background substrates are irregular and need lining to form a plumb framed surface. This surface can be fixed with plasterboard and finished with plaster or by tape and joint, which can then be decorated. Wall linings are commonly used in older buildings when the wall surface is uneven, or in buildings that are being restored in line with specific building regulations performance requirements.

When lining walls use metal, services such as electricity, water and cables can be hidden neatly behind the lining and can be accessed through service hatches.

Metal wall lining systems are used in all types of building, but are particularly suitable for those with reinforced concrete, steel frames and those that have severely uneven and out of plumb background substrates and poor **thermal insulation**.

Wall linings can be used to:
- provide fire resistance to structural steel sections within the lining cavity
- increase sound insulation
- meet thermal performance requirements of new or existing masonry walls.

KEY TERM

Thermal insulation: installed to improve the thermal performance of the building.

Benefits of a metal wall lining system

- Using a metal lining system is a cost-effective way of creating an independent wall.
- By using different plasterboards on the wall lining surface, you can achieve different performance levels to meet thermal performance, acoustic performance and vapour control requirements.
- Fixing plasterboards on the wall lining face will reduce thermal bridging, as the lining system sits slightly off the background substrate it has been fixed to.

Wall lining components

Component	Uses
Track	Floor and ceiling track for retaining the lining channel at floor, ceiling, wall, abutments and around openings.
Lining channel	This provides the main support channel to receive fixed plasterboard.
Standard bracket	Used to connect the lining channel to the structural background with a maximum 75 mm stand-off from the masonry background.

Component	Uses
Extended bracket	Used instead of a standard bracket when a maximum 125 mm stand-off is required.
Wall lining channel connector	Connectors are used to join two wall lining sections when extending their length.
Fixing channel	Used to support medium-weight fixtures on walls.
Fixing strap	Used to support horizontal board joints.
Bracket anchor	Used to fix brackets mechanically to concrete or masonry walls.
Steel framing clip	Used when encasing independent metal columns.

Component	Uses
Steel angle length	Used when forming the wall sides of metal columns attached to masonry.

ACTIVITY

Make a list of the different backgrounds you might come across on site prior to installing metal partition systems.

Wall lining junctions

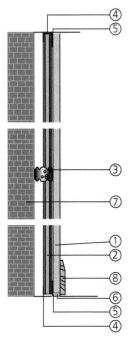

▲ Figure 7.13 Wall lining junctions: head and base

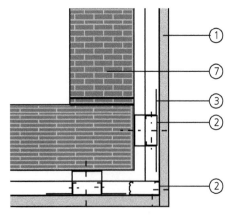

▲ Figure 7.14 Wall lining junctions: external angle

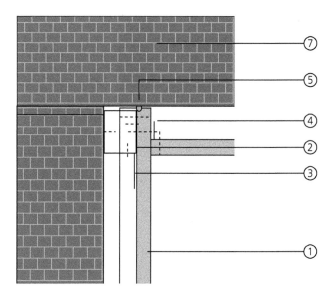

▲ Figure 7.15 Wall lining junctions: internal angle

Key:
1 Gyproc plasterboard
2 Gypframe GL1 lining channel
3 Gypframe GL2 or GL9 bracket fixed with Gypframe GL11 GypLyner anchor
4 Gypframe GL8 track
5 Gyproc sealant
6 Bulk fill with Gyproc jointing materials (where gap exceeds 5mm)
7 Wall structure
8 Skirting

Installing a wall lining system to a wall with an opening

STEP 1 Mark at each end of the wall to indicate the position of floor track and snap a chalk line. Fix the track at 600 mm centres, ensuring that the large part of the track is closest to the background.

STEP 2 Transfer and plumb the floor marks to the ceiling line with a level at each end and snap a line, then fix the ceiling track at 600 mm centres.

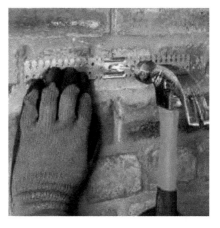

STEP 3 Mark vertical lines on the wall at 600 mm intervals to indicate bracket-fixing centres, then mark horizontal lines at 800 mm centres to show the individual bracket positions. Use a 5.5 mm drill bit to drill a 45 mm minimum depth, then position each bracket, ribs to the wall. Fix through the bracket slot into the masonry wall using an anchor fixing, which is a hammer-driven fixing.

STEP 4 Measure and cut the lining channels slightly smaller than the actual size. Round off ends with tin snips for an easier fit.

STEP 5 Bend bracket legs forwards and fix each leg to the channel using a wafer head drywall screw. Insert the screw through the hole in the bracket nearest to the back of the channel. Avoid exerting any backwards or forwards pressure on the channels when screw-fixing the brackets, otherwise a straight and true lining surface might not be achieved.

STEP 6 Bend back protruding bracket legs to sit clear of the channel face.

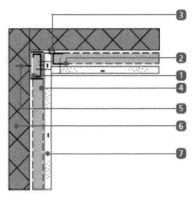

STEP 7 When fixing a wall lining at internal angles, position the lining channel tight into the corner to provide support for the lining. Bend one bracket leg across the face of the lining channel and fix with a wafer head drywall screw to secure and restrain the channel at the corner position.

STEP 8 Fix boards to all framing members at 300 mm centres using drywall screws from top to bottom. Adjacent boards are butted: ensure that screw fixings are 10 mm from bound edges and 13 mm from cut edges.

STEP 9 Adjacent linings are fixed through previous plasterboard into the frame-lining channel behind. Locate the track tight to the wall at the corner position and fix through the lining into the channel.

STEP 10 When forming an opening, position a lining channel either side of the opening to compensate for the thickness of the plasterboard to be fixed into the reveal. Then install a cut and bent track to form the head of the opening and fix to the side of the channel using two wafer head drywall screws.

STEP 11 Position short lengths of lining channel above the opening for additional support and to maintain appropriate support centres.

Metal linings can also be used to encase steel frames, columns, beams and joists, protecting them from fire and providing increased fire resistance to the structure.

IMPROVE YOUR MATHS

Calculate the cost of components required to install a wall lining system to a window wall.

ACTIVITY

1 Produce a drawing to indicate the setting out and position of frame sections, including the position of brackets.

2 With a partner, set out and install a wall lining system in line with your drawing.

Protecting a steel column using a lining system

STEP 1 Position friction-fit framing clips onto four sides of the column flanges. Position clips within 100 mm of the base and soffit and at intervals of 800 mm maximum.

STEP 2 The lining channel stand-off from the face of the structural steel frame is 25 mm and 10 mm from the edge of the flange.

STEP 3 Snap the lining channel section over the clips to form the steel framework.

STEP 4 Cut boards to width and fix to all framing members at 300 mm centres, using drywall screws. Start with a half-length board on opposite sides to **stagger** board joints around the column.

STEP 5 Cut short lengths of lining channel to form horizontal sections if you need to join adjacent boards.

KEY TERM

Stagger: where plasterboards are fixed to avoid in-line joints in ceilings and partitions.

IMPROVE YOUR MATHS

Calculate how many lining channels and brackets you need to install a wall lining system for a wall surface that measures 2.4 m in height and 2.4 m in length.

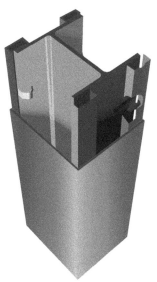

▲ Figure 7.16a Fixing wall linings to columns

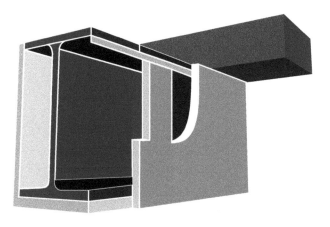

▲ Figure 7.16b Beam encasement

⑥ METAL FURRING CEILING SYSTEMS

Metal furring is a suspended ceiling system suitable for most internal dry lining applications.

The grid is fully concealed and the ceiling plasterboard lining is taped and jointed or plastered, to present a finished appearance for decoration.

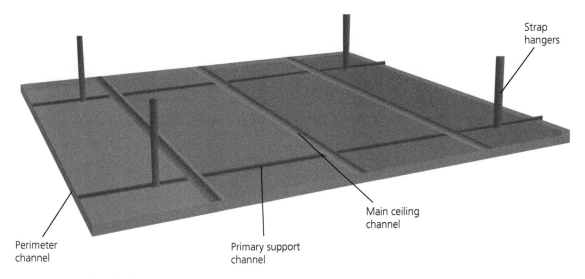

Strap hangers

Main ceiling channel

Perimeter channel

Primary support channel

▲ Figure 7.17 Metal furring system

A **bulkhead** is used to divide one ceiling from another. Usually, the ceilings have differing heights.

There are many benefits to using a metal furring system:
- Partitions or wall linings can be fixed and supported by the ceiling framework.
- Heights and gradients can be preplanned during the design stage.
- Performance plasterboards can be fixed to the framework to increase the performance level of the ceiling.
- Service hatches can be installed to allow access to the void between the metal framework and the ceiling above, if access is required for services.

▲ Figure 7.18 Bulkhead

ACTIVITY

1 Search the internet for further information on improving a building's performance rating.
2 Make a list of five **performance-enhancing** plasterboards.

KEY TERMS

Bulkhead: a partition between two compartments.

Performance-enhancing: able to improve a building's performance rating; for example, to reduce the spread of fire, minimise sound transmittance and improve thermal performance.

Metal furring ceiling system components

Component	Use
Main furring ceiling section 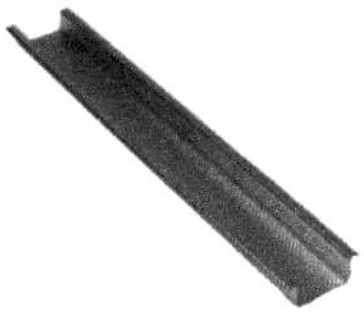	The main ceiling channel, also known as a 'top hat', for receiving the plasterboard. Provides the main support section.
Perimeter channel 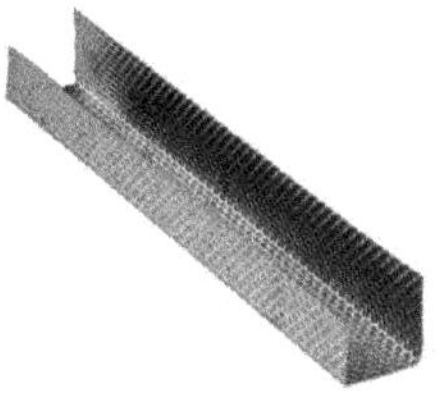	This component is fixed to the perimeter as support for the main metal furring section above.
Primary support channel 	Provides support for the main ceiling grid that receives the plasterboard lining.
Strap hanger 	Supports the suspended grid from the ceiling substrate.
Steel angle 	Can be used to form a right angle against metal surfaces and masonry surfaces.
Acoustic hanger 	Improves the acoustic properties of suspended ceilings.
Connecting clip 	Connects the primary grid and ceiling sections.

Component	Use
Nut and bolt	Used to fix the strap hanger to the soffit cleat.
Soffit cleat	Used as a suspension point from the **structural soffit**.

Installing the metal furring ceiling system

STEP 1 Before you start, you will need to determine the ceiling height. Measure from the set datum, then mark and fix the perimeter channel on the walls at 600 mm centres, using appropriate fixings. Mark fixing points for soffit cleats to the structure at 1200 mm centres to form a 1200 mm × 1200 mm grid. Secure each cleat using an appropriate fixing. You can pre-cut strap hangers or steel angle to the approximate depth of suspension required. Pre-punch or pre-drill these hangers or angles to facilitate fixing to the soffit cleats.

STEP 2 Locate each strap hanger or angle section against a soffit cleat and fix using a nut and bolt, then screw-fix to the structure. Alternatively, a steel angle can be cut, bent and drilled to facilitate direct fixing to the structure (maximum loads will be reduced by 25% if this method is used).

STEP 3 Begin to form the primary grid by fixing the first support channel. Rest one end on the top flange of the perimeter channel.

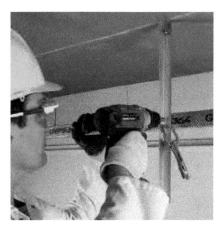

STEP 4 Fix hangers (two fixings per hanger) to the primary support channel using wafer head jack-point screws.

STEP 5 Extend channels by overlapping back to back by 150 mm minimum. Fix together using two wafer head jack-point screws.

STEP 6 Form the secondary grid by running the ceiling section at right angles to the underside of the primary grid at maximum 450 mm centres, engaging into the perimeter channel at the perimeter.

STEP 7 Alternatively, connect using connecting clips.

STEP 8 Extend sections (overlapping by 150 mm minimum) and crimp or screw-fix twice through each flange.

STEP 9 Install further sections to complete the grid. Once complete, you can fix your plasterboard.

IMPROVE YOUR MATHS

Calculate the cost of components required to install a metal furring ceiling system in a plastering bay in the workshop.

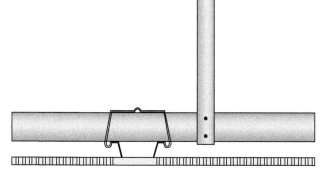

▲ Figure 7.19 A section detail for a metal furring ceiling system

ACTIVITY

1 Produce a drawing to indicate the setting out and position of frame sections in a 3 m × 3.5 m bay, including an **access panel**.

2 With a partner, set out and install a metal furring ceiling system in line with your drawing.

KEY TERM

Access panel: designed to allow easy access to areas behind walls or ceilings, to carry out maintenance of services such as cables and pipework.

IMPROVE YOUR MATHS

You need to fix a perimeter channel to a room measuring 3.5 m × 4 m. Work out and calculate the required quantity of channel if one channel measures 2.4 m.

Storing metal components

Long lengths of metal stud, track and furrings are generally stored on timber bearers, strapped and bound to avoid separation during lifting and transporting.

▲ Figure 7.20 Storing metal components

Test your knowledge

1 What type of fixing is used when securing metal track to a timber background?

A masonry nails

B dry wall screws

C galvanised nails

D wafer head screws

2 What are the maximum centres of metal stud in a metal partition wall?

A 300 mm

B 400 mm

C 500 mm

D 600 mm

3 What hand tool can be used to fix metal studs to the track?

A gauger

B ratchet

C crimper

D clamp

4 Why are deflection heads used when installing metal stud partitions?

A to control moisture

B to minimise vibration

C to allow for movement

D to prevent air leakage

5 What tool is used to ensure floor and ceiling tracks are fixed in line and accurate before fixing?

A spirit level

B metre rule

C datum gauge

D laser level

6 What metal component is used to form a window head in a metal stud partition?

A plate

B track

C stud

D bracket

7 What can be used to support plasterboards horizontally on a metal partition surface?

A strap

B plate

C anchor

D angle

8 When are clips used to install wall linings?

A when forming bulkheads

B when forming encasements

C when forming deflection heads

D when forming window openings

9 Why would this component be used with wall lining systems?

A to minimise metal vibration

B to attach linings to masonry

C to extend the track length

D to extend the wall lining height

10 In a metal furring ceiling system, what is this clip used for?

A to connect plasterboard to the ceiling channel

B to connect the primary channel to the ceiling channel

C to connect the primary channel to masonry

D to connect the ceiling channel to the perimeter

11 What type of ceiling component is shown in the image?

A masonry

B acoustic

C dry wall

D anchor

12 How are metal furrings for ceilings measured and cut?

A by width

B by length

C by height

D by depth

13 When forming door openings in metal stud partitions, what should be incorporated within the stud channel to receive screw fixings when installing door frames?

A plasterboard

B steel angle

C timber fillets

D service plugs

14 When setting out for ceiling systems, where are perimeter levels marked and transferred from?

A window reveal

B floor surface

C door opening

D datum point

15 When extending primary ceiling channels, what minimum dimensions must they be extended by?

A 100 mm

B 110 mm

C 130 mm

D 150 mm

16 What is the first component to be used when installing metal grid ceiling systems?

A secondary channel

B perimeter channel

C steel angle connector

D connecting clip

17 When setting out for installing metal wall linings, what must be the height of the fixing brackets from the floor level?

A 200 mm

B 400 mm

C 600 mm

D 800 mm

18 When setting out for installing metal wall linings, what centres should be used?

A 600 mm

B 500 mm

C 400 mm

D 300 mm

19 What document identifies the manufacturer to be used when costing metal ceiling installation systems?

A drawing

B schedule

C specification

D job sheet

20 What document provides the positions and dimensions for installing partitions?

A drawing

B specification

C job sheet

D method statement

Test your knowledge answers

CHAPTER 1

1	B	6	B
2	A	7	C
3	A	8	C
4	A	9	C
5	D	10	C

Answers to activities
Improve your maths, page 43
24.94 m²

Improve your maths, page 44
Current VAT rates @ 20%

1	60	3	168
2	96	4	216

Activity, page 44
£28.80

CHAPTER 2

1	B	11	C
2	B	12	B
3	A	13	A
4	A	14	A
5	D	15	B
6	B	16	C
7	D	17	D
8	A	18	B
9	C	19	C
10	C	20	A

CHAPTER 3

1	A	5	A
2	B	6	C
3	C	7	A
4	D	8	C

9	D	15	C
10	A	16	D
11	B	17	B
12	A	18	D
13	C	19	D
14	D	20	B

Answers to activities
Improve your maths, page 115
1. 27 screws
2. 9 sheets

Improve your maths, page 124
8 buckets of lime and 8 buckets of cement

Improve your maths, page 137
12 angle beads

CHAPTER 4

1	D	11	C
2	A	12	A
3	C	13	B
4	A	14	B
5	C	15	A
6	B	16	B
7	C	17	D
8	A	18	B
9	A	19	B
10	D	20	D

Answers to activities
Improve your maths, page 164
13 bags of sand and 4 bags of cement

Activity, page 194

1	34.44 m	3	29.35 m
2	5.09 m	4	33.36 m

CHAPTER 5

1	A	11	D
2	C	12	B
3	D	13	B
4	B	14	C
5	A	15	A
6	B	16	A
7	A	17	C
8	D	18	C
9	C	19	A
10	A	20	B

Answers to activities

Improve your maths, page 217

1.44375 m³ (round up to 1.5 m³ for delivery and order purposes)

Improve your maths, page 221

200 mm

CHAPTER 6

1	A	13	B
2	A	14	C
3	D	15	D
4	D	16	C
5	D	17	A
6	C	18	B
7	C	19	A
8	C	20	A
9	B	21	D
10	B	22	A
11	A	23	C
12	D	24	C

Answers to activities

Improve your maths, page 256

1 Eight weeks
2 £17.50

Improve your maths, page 257

1 6 lengths
2 £117.20

Improve your maths, page 259

1 3.14
2 π
3 319.9 mm

Improve your maths, page 273

A 507 minutes = 8 hours and 44 minutes
B £333.77

Improve your maths, page 276

Two trips

CHAPTER 7

1	B	11	B
2	D	12	B
3	C	13	C
4	C	14	D
5	D	15	D
6	B	16	C
7	A	17	D
8	B	18	A
9	D	19	C
10	B	20	A

Answers to activities

Improve your maths, page 291

11 lengths of C stud

Improve your maths, page 297

5 lengths of wall lining channel,
10 wall lining brackets

Improve your maths, page 302

3.5 + 3.5 + 4 + 4 = 6.25. Rounded up to 7 lengths of channel

Glossary

Access panel: designed to allow easy access to areas behind walls or ceilings, to carry out maintenance of services such as cables and pipework.

Additive: a substance added to plaster mixes to change their natural properties.

Adheres: sticks.

Aggregate: a material made from fragments or particles loosely compacted together. It gives volume, stability and resistance to wear or erosion. Coarse- to medium-grained material used in construction for bulking.

Architrave: a decorative moulding often made from timber (sometimes plaster) that is fitted around doors and windows to hide the gaps between frames and walls. It also provides a decorative feature.

Arris: the external edge of the bead.

Arris edge: a corner feature formed as a sharp edge finish with angle bead trim or by forming a hard angle.

Background key: the background surface; you may need to form (or 'key') a compatible surface to allow adhesion between various coats of plaster/render, either by using SBR/ PVA application, or by forming a scratched, rough surface between application coats to enable the coats to adhere without delaminating.

Banding: horizontal detail formed around a building at strategic points.

Batching lime mortar: mixing mortar in preparation for the following day.

BBA: within the construction industry, British Board of Agrément certification indicates a high quality, experienced and reliable company or product. It is highly regarded and used by manufacturers in industry as a symbol of quality.

Beam and block: floors made up of standard building blocks laid between pre-stressed concrete beams.

Bearers: small blocks of wood placed between and underneath materials to keep them separate and promote drying.

Bearing capacity: the ability of soil to support loads applied to the ground, so as not to produce shear failure.

Bell cast: a plaster feature set above openings and the DPC line to form a drip and deflect rain water away from the wall surface.

Binder: a material used to make the aggregate stick together when mixed.

Biological agents: types of bacteria, virus, protozoan, parasite or fungus.

Block plan: drawing that shows the proposed development in relation to the surrounding properties.

Bond [in masonry]: the arrangement of bricks or other building units when building a wall to make sure it is stable and strong. Different types of bond can be used to give a decorative effect.

Bonding agent: a substance applied to improve adhesion on poorly keyed backgrounds.

Bore hole: a narrow shaft bored (drilled) in the ground vertically or horizontally, to test soil conditions.

Bow: bend.

Brace: fixed to the top of the horse and stock at a 45° angle; it stops the running mould from twisting when it is in use.

Braced panels: critical elements of a wood/metal-framed structure which resist forces that act along the wall plane (mainly to resist lateral wind forces).

British Standards: standards produced by the British Standards Industry (BSI) Group, which is incorporated under a royal charter and formally designated as the National Standards Body (NSB) for the UK.

Building regulations: rules enforced by the building control departments of local councils to ensure all buildings are safe and fit to live and work in. These regulations contain the minimum standards for design, construction and alterations to buildings.

Bulkhead: a partition between two compartments.

Bulking: the swelling of sand when it is wet, making it heavier.

Burrs: rough edges on the profile left after filing.

Butter coat: the top coat render mixed to a buttery consistency that is applied to receive dry or wet dash finish onto its surface.

C stud: named after the profile shape of the metal stud.

Carbonation: exposure to the air to start the setting process.

Carcinogens: substances which can cause cancer.

Catalyst: commonly used with fibreglass materials or cold pour rubber, the catalyst in liquid form is carefully measured into the bulking liquid. It reacts with the other liquid: in fibreglass it hardens, whereas in cold pour rubber it turns the liquid into a flexible rubber.

Ceiling joists: horizontal components of a ceiling, to which the rafters are attached.

Chase: a void for installing services which will need to be made good with plaster.

Chattering marks: marks caused by chattering (excessive expansion of casting plaster as the mould is run along the bench when producing moulding).

Cill (sill): horizontal slat detail, forming the base of a window.

Compaction: consolidation of the sand and cement screed by tamping the screed with a box rule and floor laying trowel. This strengthens the floor screed.

Compressive load: a weight which tends to shorten or 'squash' a structure.

Compressive strength: the ability of a material (for example, insulation) to take heavy loads, such as furniture and people, without denting or going out of shape.

Consolidate: to close in the surface of a floating coat, render or floor screed with a float, making the surface flat, dense and compact.

Contamination: when materials have been in contact with something unclean, such as leaves blown into the sand or dirty water used for mixing.

Core: old moulding or plasterboard incorporated in the reverse moulding; used to reduce expansion because it reduces the amount of plaster required.

Cottage finish: a traditional render finish applied in a rustic fashion.

Craze crack: when fine cracks appear on applied plaster, caused by excessive suction in background surfaces.

CSCS (Construction Skills Certification Scheme) card: provides proof that individuals working on construction sites have the appropriate training and qualifications for their on-site job role.

Curing: allowing the mix to set and reach its full strength.

Cutting back: removing undercoat plaster from around door frames or beads, allowing you to apply setting plaster flush and preventing plaster from gathering and building up over beads.

Dabs: dry wall adhesive applied to the background to receive direct bond plasterboard installation.

Damp proof course (DPC): a layer or strip of watertight material placed in a joint of a wall to prevent the passage of water. Fixed at a minimum of 150 mm above finished ground level.

Dappling bar: used to help level and finish the liquid screed.

Datum point/line: a point or line from which measurements are taken to establish the finished floor level; usually about a metre high and running throughout the whole building.

Day work joint: used when laying large floor areas. Expansion strips are located at the edges of work completed that day. The expansion strips help to prevent cracking across the screed.

Dead man prop: a telescopic pole with pads on each end. The pole is adjusted to hold an item above your head just like an extra pair of hands.

Deep foundations: a type of foundation that transfers building loads to a subsurface layer or a range of depths.

Deflocculating additive: a substance added to a mixture, to give a slurry that would otherwise be very thick and gooey a thin, pourable consistency.

Delamination: when plaster or render becomes detached from a background and falls off, due to the eggshell effect.

Denailing: removing old nails in timber stud and timber joist backgrounds before re-installation of plasterboard.

Depth: the distance from the ceiling to the lowest edge of the cornice.

Detailed drawing: a drawing that shows the profile or design of plaster moulding in negative or positive view.

Deteriorate: become damaged, defective and unusable.

Direct bond fixing: using dry wall adhesive to fix plasterboards to solid backgrounds.

Dissipate: disappear.

Dividing wall: a wall that separates areas; for example, for framing out bedrooms and bathrooms. This will not be a load-bearing wall.

DOT: stands for Department of Transport; transported materials are given a DOT rating to indicate how hazardous they are.

Drawings: provide a graphic illustration/representation of what is to be built.

Dubbing out: the application of several coats of plaster/render to achieve a greater thickness. Each coat is applied no more than 10 mm thick, allowing for setting between coats.

Edging foam: this measures 8 mm × 150 mm × 50 m and is used to line the perimeter of each room where either a high floor build-up or a screed layer is required. The foam is used to butt all floor layers to reduce the effect of impacts transferring into the adjacent walls.

Efflorescence: a white powdery deposit on the surface of plaster, containing a high proportion of salt.

Eggshell: when plaster or render dries out too quickly, shrinks and cracks.

Elevations: drawings that show the external walls of the building from different views.

Energy certificate: states a property's energy efficiency and recommends how energy can be saved, to save money and be environmentally friendly.

Enrichment: decorative sections of plasterwork, such as egg and dart or acanthus leaves.

Environment Agency (EA): public body working to protect and improve the environment.

Expanded metal lath (EML): sheet material in the form of diamond-shaped mesh that is used to reinforce a surface. This material can be fixed with screws and plugs or galvanised nails, or it can be bedded into the render material.

External slurry: thin, sloppy mixture of cement and bonding adhesive applied to a background to bond render to its surface.

Extruded polystyrene: this is formed by heating polystyrene crystals to high temperatures, along with other additives, and forcing the mixture through a die (which is like a mould). The result is a denser material than expanded foam.

Faced/Fair-faced: these bricks are durable and graded on a scale to match the building material required by the project. Cosmetic face bricks are made to face the world with a smooth look, whereas common bricks/blocks do not have smooth sides.

Fencing: ensuring that a model such as an enrichment is surrounded, to eliminate leakage when flexible compounds are poured over the enrichment model to form a reverse.

Fibre hand: a person who manufactures and installs fibrous plaster components.

Filling out: building out an uneven background.

Fire proofing: installed to increase the fire rating of the metal frame.

First fix: all work (carpentry, electrical or plumbing) carried out before plaster is put on internal walls.

Flexural strength: the ability of a material to bend without cracking or breaking.

Foot traffic: people walking or travelling over an area. The term is used when a newly laid floor screed allows someone to walk over the screed without leaving any indentations, usually after about three or four days.

Foreshortening: shows an object or view as closer than it is; dramatically reduces an object in scale.

Forest Stewardship Council (FSC): an international, non-governmental organisation dedicated to promoting the responsible management of the world's forests.

Frame members: studs/partitions, wall plates and lintels.

Furrings: metal stud wall or ceiling linings, fixed with plasterboard; also known as metal-framed backgrounds.

Gable apex: the triangular part of a gable wall.

Galls: blemishes in a plaster surface due to poor workmanship.

Galvanised: coated with zinc to prevent corrosion.

Gauge: the term used to indicate the metal's thickness.

Gauging: measuring out the ratio for mixing materials.

Green: describes a material such as concrete/screed/plaster that has not fully set and is still soft.

Grinning: when the plaster surface reveals imperfections caused by deeply keyed devil floating or variable suction of the background.

Gross cost: value of something including taxes and other costs.

Gypsum: soft sulphate mineral widely mined and used in many types of plaster, available in fine-grained white or lightly tinted varieties.

High-density block: durable and resilient, high in strength and with good acoustic rating, generally used for structural purposes.

Hoarding: a barrier surrounding a site to protect against theft and unauthorised entry.

Hollowness: holes or depressions in previously plastered walls, where the plaster has become loose from the background.

Horse: runs against the running rule on the bench.

Improvement notice: issued by the HSE or local authority inspector to formally inform a company that safety improvements are needed.

In situ: when a plaster moulding is run directly to the background, using a positive profile.

Insertion: incorporating an enrichment pattern in to a moulding to enhance its design.

Insulation: objects or materials used in buildings to improve thermal quality.

Jamb: when the track is cut and returned to form a door or window opening in a partition.

Key: referring to the background surface. A rough surface produces adequate key; smooth surfaces have less or no key.

Key stone: detail at the apex of a formed arch.

Keyed surface: a surface able to receive an application of plaster/render that enables suitable adhesion of two surfaces.

Kyoto Protocol: an agreement between the world's nations to reduce greenhouse gas emissions.

Laitance: a layer of weak cement that can affect the strength of the floor screed if not removed.

Lapped: the overlap of material, such as DPM, to ensure no moisture can penetrate.

Lateral load: typical lateral loads include wind blowing against a facade, an earthquake, or ocean movement on beach-front properties.

Legislation: a law or set of laws suggested by a government and made official by a parliament.

Lime blooming: this happens when lime in the form of calcium hydroxide migrates and forms on the surface as the material dries out. On reaching the surface, this reacts with carbon dioxide in the air and produces a surface deposit of calcium carbonate.

Linear measurement: measurement of a straight distance between two points.

Linings: timber surround for internal doors, forming a lining to the masonry or studded opening.

Listed building: a building of particular interest, architecturally or historically, which is considered to be of national importance; details of these buildings are recorded on national lists.

Longitudinal: running lengthwise rather than across.

Low-density block: extremely versatile and can be used in standard wall construction.

Manufacturer's instructions: these state what a product may be used for, how it is to be installed and the conditions it can safely be exposed to.

Manufacturer's technical information (MTI): providing technical information on products for safe use and correct installation.

Mechanical fixings: fixings used to fix EML to composite backgrounds.

Metal ceiling lining: a metal grid background to form accurate, level ceiling systems in old and modern buildings.

Method statement: a document to help manage work and ensure that everyone has been told about taking precautions. It often includes a logical sequence of work.

Metric scale: a system of measurement in millimetres.

Mitre: a cut joint to an internal or external angle. The joint is then made good (the gaps are filled) with casting plaster or fixing adhesive, using a joint rule and a small tool.

Movement joints: these allow surface and components to move slightly under expansion and contraction in buildings.

Muffle: a temporary plaster mix applied to extend the profile by 5–6 mm; thin ply or a zinc sheet can also be used.

Mullion: vertical bar detail around windows.

Mutagens: agents such as radiation or chemical substances which can cause genetic mutation in the body.

Net cost: value of something after taxes and other costs have been deducted.

Noggin: a timber strut fixed between timber studwork or timber joists to strengthen and prevent twisting.

Nominal: standard.

Packers: small pieces of doubled-up offcut that the boards can sit on to keep them off the floor.

Partition: wall used to separate and divide the overall space within a building into rooms.

Party wall: a dividing partition between two adjoining buildings that is shared by the occupants of each residence or business.

Passive housing: creating an ultra-low energy building with a small ecological footprint that requires little energy for space heating or cooling.

Penetrating damp: moisture travelling through the wall from outside.

Performance-enhancing: able to improve a building's performance rating; for example, to reduce the spread of fire, minimise sound transmittance and improve thermal performance.

Pick up: when the materials start to stiffen.

Pilot hole: a small, pre-drilled hole bored to help prevent splitting.

Planar wall: a flat wall.

Plant: machinery, equipment and apparatus used for an industrial activity. In construction, plant refers to heavy machinery and equipment used during construction works such as diggers, dumpers and cranes.

Plasterboard strut: used to prop the plasterboard in position prior to securing with screws.

Plinth: the surface area below the bell cast that runs along the DPC; upstand detail formed at the bottom of a building at ground level.

Polymer: strong glue-like substance used to improve the adhesion of render surfaces.

Popping: where plaster comes away from the plasterboard background

because a fixing is loose or has been driven too far into the plasterboard surface.

Porosity: how porous a material is; a porous material has many tiny holes or 'pores' in it and will easy absorb air or liquid.

Portland Cement: also known as Ordinary Portland Cement (OPC), this is the most commonly used cement. It is named after stone quarried on the Isle of Portland off the British coast, as it is similar in colour.

Profile: the shape and pattern of a mould outline.

Prohibition notice: issued by the HSE or local authority inspector when there is an immediate risk of personal injury. This is very serious and a company that receives a prohibition notice will clearly be breaking health and safety regulations.

Project brief: a summary of a project's ideas; it shows what work needs to be done.

Projection: the distance from the wall to the outer edge of the top of the cornice.

Pull in: stiffen up or start to set.

Purlins: roof-framing members that span parallel to the building eaves and support the roofing materials.

Quoin: detail formed at external corners of a building.

Ratio: the proportion of materials mixed together; for example, 6 parts of sand to 1 part of cement would be written as 6 : 1.

Renovation work: repairs to old, deteriorated buildings that need to be upgraded and modernised or restored to their original state.

Restoration work: restoring plasterwork back to its original state.

Retarder: a chemical additive that slows down the setting time of gypsum plasters, casting plaster and cement.

Return: a corner profile/edge.

Reveal: small return to a window or door opening.

Rising damp: moisture rising up from the floor through the wall.

Risk assessment: the process of identifying hazards and risks that could cause harm.

Rotating: small circular movement when applying brush textured finish.

Rule: flatten off plaster/render using an aluminium darby/straight edge rule.

Run cast: a plaster moulding run on a bench with an upstand to produce a positive profile.

Scabbled: roughened.

Scabbling: removing the surface finish by mechanical means, producing a suitable key.

SCAFFTAG: a scaffold-status tagging system to prevent hazards when working at height and efficiently manage the inspection procedures for scaffolding.

Scale ratio: the ratio of the size of a drawing to the size of the object being drawn.

Schedule: a timetable or sequence of events.

Scratch coat: a plaster or render mix applied to a surface to control suction and provide adequate key before a floating coat is applied.

Screed levelling tripod: used to make sure the floor is laid to the correct level.

Scrim: used to reinforce plasterboard butted joints to reduce cracking before applying finishing plaster.

Seismic forces: forces which act on a building to represent the effect of an earthquake.

Seismic load: relates to forces caused by ground movement such as earthquakes, which will cause movement and possible collapse of structures.

Semi-dry: the mix consistency of a traditional sand and cement screed.

Service channels: gaps or channels incorporated in the studs to allow for cables and piping.

Shallow foundations: a type of foundation that transfers building loads to the earth very near to the surface.

Shear failure: occurs when there is not enough resistance between materials, so structures can move and flex; this leads to structural weakness and cracking.

Shelf life: the use-by date of products such as cement and lime.

Shrinkage: applied plaster can shrink as it dries out, forming small cracks.

Site manager: responsible for the completion of a building project effectively, safely and on time.

Site plan: shows the plot in more detail, with drain runs, road layouts and size and position of existing buildings.

Skim: the term used by some plasterers to describe the setting coat.

Skirting board: a decorative moulding often made from timber (sometimes plaster) that is fitted at the bottom of a wall to hide the gap between wall and floor and to protect the bottom of a wall from foot traffic.

Sleeper wall: a short wall used to support floor joists, beam and block or hollowcore/concrete slabs at ground level.

Slumping: when plaster has been applied too thickly and slides down the wall due to excessive thickness and weight.

Slurry: a wet mix applied with a brush.

Snots: residue left on the surface of the floating coat after consolidation. This must be removed to prevent it from penetrating the surface of the setting coat.

Soffit: the underside of a window or door opening.

Sole trader: a self-employed person who owns and runs their own business. The business does not have a legal identity separate to its owner, so that person is the business.

Sound proofing: installed to reduce noise transfer through metal systems.

Specification: instructions stating the standards required and practice to be followed for a task, usually BBA-approved and to meet British Standards. It is often an official document from the architect who is overseeing a project.

Spotting: applying a small amount of jointing filler over penetrated fixings.

Squeeze: a method for reproducing a mould outline.

Stagger: where plasterboards are fixed to avoid in-line joints in ceilings and partitions.

Stakeholder: a person with an interest or concern in a project, especially business.

Stock: holds the zinc profile and is attached at a 90° angle to the horse.

Stock rotation: ensuring old stock is used before new stock. When new stock is delivered, it should be stored behind the older stock, which needs to be used first.

Strike off: the built-up plaster area on the back of a cast that will come into contact with the background surface on the wall and ceiling when the plaster cornice is fitted in place.

Structural soffit: the background from which the system is suspended; for example timber joist, concrete, beam and block.

Strut/prop: a telescopic pole with pads on each end; the pole is adjusted to hold an item above your head, just like an extra pair of hands. A useful piece of equipment when working on your own.

Subcontractor: a tradesperson who is not directly employed by a company but is employed for short periods to complete some aspects of the work. They are paid for the completed work at a set price.

Subsoil: the layer of soil under the topsoil on the surface of the ground. It is composed of a variable mixture of small particles such as sand, silt and clay, but with a much lower percentage of organic matter and humus (a dark, organic material that forms in soil from plant and animal matter decay).

Suction: the rate at which a background absorbs moisture.

Tamped finish: where the screed has been compacted and consolidated to push coarse aggregate below the screed surface.

Tender: to submit a cost or price for work in an attempt to win the contract.

Thermal insulation: installed to improve the thermal performance of the building.

Thermoplastic: a characteristic of material, meaning it can be remelted.

Thixotropic: material that remains in a liquid state in its container, but changes to a gel-like state and hardens into position when brushed vertically or horizontally.

Three coat work: when plastering exteriors, this means applying three distinct layers of render: dubbing out/pricking up, scratch and finish render surface.

Timber rule: straight plane timber used as a guide to form the edge of a return. Before the introduction of aluminium feather edges, timber rules were also used as straight edges.

Tolerance: required standards and accuracy of completed work.

Topographic survey: a survey that gathers data about elevation points on a piece of land and presents them as contour lines on the plot. It gives information about the natural and human-made features of the land, such as natural streams or existing groundworks.

True: accurate to plumb, level and/or line.

U-values: a measure of heat loss through a building's walls, floors and roof. Higher U-values suggest poor thermal performance. The lower the U-value, the better the building is at retaining heat.

Vertical load: loads in addition to the weight of the structure; this can include the weight of floors, roofs, beams and columns all pushing down compressively.

Viscosity: how thick or runny a liquid is. A viscous liquid is thick and sticky and does not flow easily.

Voids: pockets of air, common in poorly graded sand.

Volatile organic compounds (VOCs): organic chemicals that have a high vapour pressure at room temperature, including human-made and naturally occurring chemical compounds. Nearly all scents and odours are classed as VOCs.

Wet screed: band of undercoat plaster screed used as a floating guide while still wet.

Zinc: a non-ferrous metal, zinc is easy to cut and forms its own protective layer called patina, so it does not rust.

Index

Notes